Adaptive Control Theory and Applications

Edited by Petros Ioannou

Published in London, United Kingdom

Adaptive Control Theory and Applications
http://dx.doi.org/10.5772/intechopen.1001654
Edited by Petros Ioannou

Contributors
Ali Turker Kutay, Andrea L'Afflitto, Andreja Goršek, Anthony J. Calise, Azwirman Gusrialdi, Bozenna Pasik-Duncan, Boštjan Polajžer, Darius Andriukaitis, Darja Pečar, Deepalakshmi Babu Venkateswaran, Francisco Jurado, Francis F. Assadian, Giri M. Kumar, Jožef Ritonja, Mattia Gramuglia, Metehan Yayla, Nhan T. Nguyen, Petros Ioannou, Quang M. Lam, Rongfei Li, Tyrone E. Duncan, Zhihua Qu

First published in London, United Kingdom, 2024 by IntechOpen
IntechOpen is the global imprint of INTECHOPEN LIMITED, registered in England and Wales, registration number: 11086078, 167-169 Great Portland Street, London, W1W 5PF, United Kingdom

For EU product safety concerns: IN TECH d.o.o., Prolaz Marije Krucifikse Kozulić 3, 51000 Rijeka, Croatia, info@intechopen.com or visit our website at intechopen.com.

British Library Cataloguing-in-Publication Data
A catalogue record for this book is available from the British Library

Adaptive Control Theory and Applications
Edited by Petros Ioannou
p. cm.
Print ISBN 978-0-85466-713-0
Online ISBN 978-0-85466-712-3
eBook (PDF) ISBN 978-0-85466-714-7

Meet the editor

Petros Ioannou received his Ph.D. from the University of Illinois, USA, in 1982. After receiving his Ph.D., Dr. Ioannou joined the Department of Electrical Engineering, University of Southern California, USA, where he is a professor and the A.V. "Bal" Balakrishnan Chair. He is a lifetime fellow of the Institute of Electrical and Electronics Engineering (IEEE) and a fellow of the International Federation of Automatic Control (IFAC) and the American Association for the Advancement of Science (AAAS). He is the author and co-author of 9 books and more than 400 research papers on adaptive control and transportation. He is a member of the National Academy of Engineering, a fellow of the National Academy of Inventors, and a foreign member of Academia Europaea and the European Academy of Sciences.

Contents

Preface

Adaptive control is an evolving field in both research and applications. In most applications, parametric uncertainties are inevitable due to the physics of the problem and the wear and tear experienced over time. Techniques like adaptive control help to improve performance as well as extend the life of equipment. While most control theory is developed for linear time invariant (LTI) systems, with a well-established collection of analytical and design tools, adaptive control introduces complexities. The inclusion of an online parameter estimator makes the closed-loop system nonlinear, making it less understood than traditional control approaches.

Despite advancements in theoretical and design tools that guarantee stability and robustness, defining tight bounds for stability, robustness, and performance remains challenging. These bounds depend on the parameter estimator's performance, which in turn is influenced by the quality of data it receives. Since this data is generated within the feedback loop, its adequacy in accurately estimating parameters may not converge to their values. Nevertheless, adaptive control schemes guarantee stability, robustness, and good performance, often relying on the selection of filters and design parameters, which are often selected through trial and error to improve performance.

These limitations, compared to classical control of LTI systems, are to be expected. Adaptive control relies on parameter estimation to achieve its objectives, which often involves signals that approach zero or follow specific trajectories–objectives that conflict with the parameter estimation process, which requires large, dynamic signals to excite the system for accurate parameter identification. This conflict is a well-known challenge in adaptive control. For these reasons, adaptive control is particularly useful, and sometimes the only viable option, when classical control designs cannot handle large parametric uncertainties. For these reasons, research in adaptive control continues to focus on understanding the interplay between parameter estimation and control, transient behavior, and the design of filters and parameters to improve robustness and performance.

This book presents recent results and applications of adaptive control.

The book starts with Chapter 1, "Introductory Chapter: Robust and Adaptive Control", which describes the main features of adaptive control and provides a list of some popular books on the subject.

Chapter 2, "On Advances in Noise Modeling in Stochastic Systems, Control and Adaptive Control", discusses advances in noise modeling in stochastic systems, control, and adaptive control. It includes formulations and solutions as well as presents future problems for investigation. Noise is modeled using Brownian motion, fractional Brownian motion, and Rosenblatt processes. The chapter also presents results for adaptive control of partially known linear stochastic systems with Brownian motion and fractional Brownian motion.

Chapter 3, "Information Recovery in Composite Model Reference Adaptive Control", introduces a new adaptive control approach that incorporates a frequency-limited estimation of matched uncertainties. This method incorporates a low-frequency uncertainty estimation combined with a time-varying learning rate structure. Closed-loop stability is analyzed using Lyapunov stability theory, with numerical examples and software-in-the-loop simulations using an X-plane flight simulator.

Chapter 4, "Robust Hybrid Model Reference Adaptive Control and Output-Feedback Linearization with Applications to Quadcopter UAVs" describes a robust model reference adaptive control (MRAC) system for hybrid, time-varying plants with parametric, matched, and unmatched uncertainties, as well as uncertainties in the plant's discrete-time dynamics. The proposed method is applied to control of output-feedback linearized dynamical models and addresses the problem of controlling the dynamics of a quadcopter unmanned aerial vehicle (UAV) tasked with following both a user-defined trajectory and a user-defined attitude.

Chapter 5, "Decentralized Robust Direct MRAC for the Attitude of a Quadrotor UAV", deals with the design and analysis of a decentralized robust direct model reference adaptive controller to prevent parameter drift in adaptive control of the attitude of a quadrotor UAV. The quadrotor's rotational dynamics are parameterized in a decentralized manner, with robustness strategies like dead-zone, sigma-, and e-modifications implemented.

Chapter 6, "Perspective Chapter: Intelligent Adaptive Flight Control Augmentation of a Dynamic Inversion Autopilot", employs an intelligent flight controller to augment a nonlinear dynamic inversion (NDI) autopilot. This approach helps in restoring aircraft stability and command-following capabilities when subjected to effector performance degradation or failures. The proposed adaptive control approach is compared with traditional MRAC and evaluated for its potential in high-Mach UAV applications.

Chapter 7, "Practical Approaches to the Control of Milk Fermentation with Kefir Grains", focuses on controlling milk fermentation for kefir production. It presents the design of a closed-loop control system that adjusts bioreactor temperature to regulate dissolved CO_2 concentration. Both constant-parameter control techniques and adaptive control systems are employed to track desired reference trajectories during the fermentation process.

Chapter 8, "Enhancing Resilience in Cooperative Systems against Cyber-Attacks: A Defense Framework through Adaptive Network Reconfiguration and Digital Twin", presents a solution that enhances the resilience of multi-agent cooperative systems against cyberattacks. The framework uses a distributed algorithm to identify critical edges in the network via online parameter estimation. An adaptive network reconfiguration algorithm removes these edges, ensuring resilience. Numerical simulations validate the effectiveness of the approach.

Chapter 9, "Innovative Adaptive Imaged Based Visual Servoing Control of 6 DoFs Industrial Robot Manipulators", proposes a feedforward-feedback adaptive control algorithm based on the Youla parameterization method. This approach stabilizes nonlinear camera and robot manipulator models across the full range of operations.

The adaptive algorithm linearizes and decouples the model online, ensuring robust and effective control across various industrial robotic systems. Simulation scenarios illustrate the proposed controller's robust performance.

The editor would like to thank all the authors of the contributing chapters as well as Ms. Tea Jelaca at IntechOpen for supervising the submission, monitoring the process, and providing guidance throughout this project.

Petros Ioannou
University Professor,
Department of Electrical and Computer Engineering,
University of Southern California,
Los Angeles, California, USA

Chapter 1

Introductory Chapter: Robust and Adaptive Control

Petros Ioannou

1. Introduction

The words "*adaptive systems*" and "*adaptive control*" have been used as early as 1950 [1]. Traditionally adaptive control is defined as the combination of a parameter estimator to estimate unknown parameters in real time to be used by the feedback controller for linear time invariant (LTI) systems which may change over time or due to the environment. Over the years, this definition has expanded to include neuroadaptive schemes for nonlinear systems where the parameters are the weights of the neural network. In general, any feedback or open-look control system that includes some form of parameter estimation to deal with large parametric uncertainties and parameter variations over time is often labeled as adaptive control. Since the early 80s, several books and [2–15] and many papers have been published covering different techniques and aspects of adaptive control. An adaptive control scheme consists of two parts: the learning of the parameters and the control part. The way of designing the learning mechanism and adjusting the controller characteristics in response to changes in the plant and disturbance dynamics distinguishes one scheme from another and there is a wide range of approaches to achieve that. Some are explained in the following subsection.

2. Learning the parameters

Learning the values of the parameters that are needed to calculate the controller parameters used to generate the control input in real time can take different forms. One approach is to do most of the work offline, often referred to as training, which identifies the best parameter values for the controller at each operating point or region and creates long tables with these stored values. During online operations, measured data are used to identify the operating point and get the values of the controller parameters from the look up table. This method known as *gain scheduling* has been the traditional method of adaptation applied in aerospace and other applications and is still used today in many control applications [5]. One of its drawbacks is that any unpredictable changes not captured by the gain scheduling tables cannot be accommodated.

The most popular adaptive control structures are those based on a parameter estimator which generates online estimates of the unknown parameters that are then used to calculate the controller parameters. The design of the parameter estimator relies on being able to express the unknown parameters in the form of a linear-in-parameters parametric model of the form

$$z = \theta^{*T}\Phi \tag{1}$$

where z is a signal available for measurement, such as the filtered value of output y, Φ is a vector that is known and depends on filtered values of available measurement signals, and θ^* is the unknown parameter vector. Other forms of parametric models are described in [5]. The parameter estimator based on (1) is usually of the form

$$\dot{\Theta} = \Gamma\varepsilon\Phi, \varepsilon = \hat{z} - z, \hat{z} = \theta^T\Phi \tag{2}$$

where Γ is a positive definite matrix and could be time-varying and generated by some other equation, $\hat{z}$ is the predicted value of the signal z based on the parameter estimate θ(t) at time t, and ε is the estimation error. The way we design the adaptive gain Γ and modify the form of the estimation error can lead to different parameter estimators, also referred to as adaptive laws. The above parameter estimator or adaptive law guarantees in general that the parameter estimate θ is always bounded and its derivative is square-integrable, provided the so-called regressor vector Φ is bounded but does not guarantee that θ will converge to the unknown parameter θ^* unless Φ satisfies some signal richness properties. The necessary and sufficient condition for θ to converge to the unknown parameter θ^* exponentially fast is that Φ is persistently exciting (PE) [2–15]. This PE property cannot be guaranteed in a feedback loop where the control objective may require that Φ is weak or its elements go to zero. The significant result in adaptive control is that this PE condition is not required to meet the control objective. This leads to some robustness issues as the lack of exponential convergence of the estimated parameters to the desired ones cannot guarantee bounded signal estimates in the presence of modeling errors that change the form of the linear parametric model above into

$$z = \theta^{*T}\Phi + \eta \tag{3}$$

where η is an unknown signal that depends on disturbances and unmodeled dynamics. In this case, the adaptive law (2) can no longer guarantee bounded estimates. As shown in [2, 5], a small disturbance that goes to zero asymptotically with time may cause the estimated parameters to drift to infinity. The "pure" integrator form in (2) and lack of PE property, which is the case in most adaptive control schemes, cannot guarantee a bounded input bounded state stability. One way to prevent parameter drift is to modify the pure integral action using a small feedback term known as σ-modification to obtain

$$\dot{\Theta} = -\sigma\theta + \Gamma\varepsilon\Phi, \varepsilon = \hat{z} - z, \hat{z} = \theta^T\Phi \tag{4}$$

where $\sigma > 0$ is a very small constant. Other modifications include the switching σ, parameter projection, dead zones, ε-modification analyzed and compared in [5].

Another approach to learning is to do a lot of experiments offline and develop a set of possible plant models or possible stabilizing controllers leading to similar schemes as gain scheduling.

3. Adaptive control structures

An adaptive control scheme is often defined as the feedback control system that incorporates a parameter estimator, which generates parameter estimates at each time t to be used by the controller. The form of the controller is the one used when the unknown parameters were fully known. The way the parameter estimator is combined with the control law leads to different configurations, some of which are briefly described in the following subsections. The form of the control law is designed so that if the plant parameters were known, the controller gains can be calculated. Since the plant parameters are unknown, the controller parameters are calculated at each time t by following the same approach as in the known parameter case, but instead of using the unknown plant parameters, we use their estimate generated by an adaptive law. This structure of adaptive control is known as *indirect adaptive control*. For the calculation of the controller parameters to be possible, the plant is often required to be controllable or stabilizable. However, since the estimated parameters may not correspond to a plant that is stabilizable, the calculation of the controller gains may not be possible at each time t. This is the well-known sterilizability issue [2–12]. This intermediate step of calculations can be bypassed for certain classes of plants leading to *direct adaptive control* [2–15]. Another adaptive control approach is to use offline experiments and design all possible controllers by assuming a class of different possible plant models. Then the challenge will be to identify which controller from the set of predetermined controllers is the one to use as the plant changes by using a properly designed switching logic. This leads to the class of *switching adaptive control schemes* [11, 16].

The design of adaptive control schemes for linear time-varying plants with unknown parameters is presented in [10]. These results were followed by further extensions to nonlinear systems [14, 15], where, in all cases, the parametrization of unknown parameters multiplying known signals is possible.

Author details

Petros Ioannou
University of Southern California, California, USA

*Address all correspondence to: ioannou@usc.edu

References

[1] Aseltine JA, Mancini AR, Sartune CW. A survey of adaptive control systems. IRE Transactions on Automatic Control. 1958;**3**(6):102-108

[2] Ioannou PA, Kokotovic PV. Adaptive Systems with Reduced Models: Lecture Notes in Control and Information Sciences. Vol. 47. New York: Springer-Verlag; 1983

[3] Landau ID. Adaptive Control: The Model Reference Approach. New York, Philadelphia, PA, USA: Marcel Dekker; 1979

[4] Goodwin GC, Sin KC. Adaptive Filtering Prediction and Control. New Jersey: Prentice Hall; 1984

[5] Ioannou PA, Sun J. Robust Adaptive Control. Prentice-Hall, Inc; 1996 (out of print in 2003), published by Dover, 2006

[6] Astrom KJ, Wittenmark B. Adaptive Control. Reading, Massachusetts: Addison-Wesley Publishing Company; 1989

[7] Narendra KS, Annaswamy AM. Stable Adaptive Systems. New Jersey: Prentice Hall; 1989

[8] Sastry S, Bodson M. Adaptive Control: Stability. Prentice Hall, New Jersey: Convergence and Robustness; 1989

[9] Landau R. Lozano and M. Adaptive Control, Springer: M'Saad; 1997

[10] Tsakalis KS, Ioannou PA. Linear Time Varying Systems: Control and Adaptation. New Jersey: Prentice Hall; 1993

[11] Ioannou PA, Fidan B. Adaptive Control Tutorial. Garden City, NY, USA: SIAM; 2006

[12] Tao G. Adaptive Control Design and Analysis. Hoboken, New Jersey, USA: John Wiley & Sons; 2003

[13] Anderson BDO, Bitmead RR, Johnson CR, Kokotovic PV, Kosut RL, Mareels I, et al. Stability of Adaptive Systems. Cambridge, Massachusetts: MIT Press; 1986

[14] Krstic M, Kanellakopoulos I, Kokotovic P. Nonlinear and Adaptive Control Design. Hoboken, New Jersey, USA: John Wiley & Sons; 1995

[15] Farrell J, Polycarpou M. Adaptive Approximation Based Control: Unifying Neural, Fuzzy and Traditional Adaptive Approximation Approaches, Hoboken. NJ: J. Wiley; 2006

[16] Anderson B, Brinsmead T, Liberzon D, Morse AS. Multiple model adaptive control with safe switching. International Journal of Adaptive Control and Signal Processing. 2001;**15**: 445-470

Chapter 2

On Advances in Noise Modeling in Stochastic Systems, Control and Adaptive Control

Bozenna Pasik-Duncan and Tyrone E. Duncan

Abstract

In these chapter describes various advanced models for noise in systems. Some results associated with them and the related control and adaptive control problems are formulated and solved. Some problems are proposed for future investigation. The processes used for modeling a system noise include Brownian motion, fractional Brownian motion, and Rosenblatt processes. The Rosenblatt processes are non-Gaussian processes that have long-range dependence, an important property for current applications. They can be considered as a non-Gaussian generalization of fractional Brownian motions that also have a long-range dependence property. Some results are presented for adaptive control of partially known linear stochastic systems with Brownian motion and with fractional Brownian motion. Some optimal control results are given for scalar linear stochastic systems with Rosenblatt noise.

Keywords: stochastic systems, stochastic control, stochastic adaptive control, noise modeling, non-Gaussian noise

1. Introduction

Many physical systems experience perturbations or have unmodeled dynamics in the systems. These perturbations can often be modeled by white noise or other noise processes. Examples show that noise may have a stabilizing or a destabilizing effect. Some stochastic models are considered here with different types of noise. The noise processes include Brownian motion, fractional Brownian motion, and Rosenblatt processes. These latter two processes often seem more appropriate in a model given the data than using a Brownian motion. All three noise processes are used to drive linear and semilinear systems. However, each type of noise process for a control system needs somewhat distinct methods of analysis for solutions to linear-quadratic control and adaptive control problems. The general approach to adaptive control that is described here exhibits a splitting or separation of identification and adaptive control. Adaptive control results are given for partially known linear and semilinear systems with Brownian motion [1] and linear and semilinear systems with fractional Brownian motion [2–5]. The solution of an ergodic control problem for a scalar linear system with a Rosenblatt noise is also given [6]. Stochastic calculus methods were developed for fractional Brownian motion [7–10] and for Rosenblatt processes [11]. Stochastic calculus provides powerful tools: stochastic integrals, Itô's differential formula, and martingales.

2. Noise modeled by Brownian motion

Historically, noise in continuous-time physical systems has been modeled by Brownian motion. This Gaussian process was a natural choice from the Central Limit Theorem and was used for the early applications by Kolmogorov and Wiener. Furthermore, a stochastic calculus had been developed particularly by K. Itô to allow for the analysis of the continuous-time stochastic systems. The system models were typically ordinary differential equations driven by a Brownian motion denoted as stochastic differential equations. Since the stochastic systems for these problems were typically only partially known, there was a problem to identify the appropriate values for the unknown parameters in the equations. These problems are typically described as identification problems for stochastic systems. The linearity of the systems provided explicit solutions and various testing methods were used, such as least-squares or weighted least-squares estimation, to obtain the estimates of unknown parameters in the systems. A general model is given as

$$\begin{aligned} dX(t) &= AX(t)dt + BU(t) + CdW(t) \\ X(0) &= x_0 \end{aligned} \tag{1}$$

where $X(t) \in R^n$ and $W(t)$ is a Brownian motion. Assuming that the matrices A, B, C are unknown or at least only partially known, there is a problem with identification of these unknowns. A typical method has been the least-squares method, which can provide explicit and computationally feasible algorithms for the estimates. In particular, the estimates are least-squares parameter estimation for the matrices A, B. An estimate for the matrix C can be obtained from the quadratic variation of the solution $X(t)$, which means the quadratic variation of the stochastic term $(CdW(t))$ that uses the quadratic variation of a Brownian motion.

The cost functional for the adaptive control problem is

$$J(U) = \limsup_{T\to\infty} \frac{1}{T}\int_0^T (X^T(t)Q_1X(t) + +U^T(t)Q_2U(t))dt \tag{2}$$

where Q_1, Q_2 and symmetric and positive definite matrices. A weighted least-squares estimation procedure is done for $\theta^T = [A \quad B]$ and ϕ is the vector formed from $X(t)$ and $U(t)$.

The equation for X can be rewritten as

$$dX(t) = \theta^T\phi(t)dt + CdW(t) \tag{3}$$

where the weighted least-squares estimation scheme is given by the following equations

$$\begin{aligned} d\theta(t) &= a(t)P(t)\phi(t)\left[dX^T(t) - \phi^T(t)\theta(t)dt\right] \\ dP(t) &= -a(t)P(t)\phi(t)\phi^T(t)P^T(t)P(t)dt \end{aligned} \tag{4}$$

It can be shown that these processes form a family of strongly consistent estimators meaning that the estimators of the unknown terms A, B converge almost surely [1]. It follows that a family of certainty equivalence controls is self-optimizing, that is, they provide convergence to the optimal long-run average cost for the true system. This approach eliminated some other assumptions that have previously been used that are unnecessary for the control problem for a known system and are often difficult to verify. For the adaptive control, a diminishing excited certainty equivalence control is used. To obtain a strong consistency for the family of estimators, a diminishing excitation is added to the adaptive control. The complete solution to the adaptive control problem with the most natural assumptions has been obtained for continuous-time linear and some nonlinear systems [1].

3. Noise modeled by a fractional Brownian motion for a linear system

Since many physical phenomena have a long-range dependence, it is natural to consider a Gaussian process with a long-range dependence property, that is, a fractional Brownian motion, and then to consider adaptive control for a linear stochastic equation driven by a fractional Brownian motion. If the equation has unknown parameters, then one has a problem of adaptive control. This problem is described here for a scalar linear system driven by a fractional Brownian motion. These processes are indexed by the Hurst parameter $H \in (0, 1)$. Initially, a fractional Brownian motion is defined to fix the notation.

Fractional Brownian motion (FBM) is a collection of Gaussian processes with continuous sample paths and a stochastic calculus [7, 8, 10, 12] whose definition is given now. These processes and stochastic calculus tools had been developed for the purpose of being applied to linear and semilinear control systems with some unknowns [7, 8].

Definition 3.1. For each $H \in (0, 1)$, a real-valued standard fractional Brownian motion $(B(t), t \geq 0)$ is a Gaussian process with continuous sample paths that satisfies the following conditions:

$$\mathbb{E}[B(t)] = 0 \tag{5}$$

$$\mathbb{E}[B(s)B(t)] = (1/2)\left(t^{2H} + s^{2H} - |t - s|^{2H}\right) \tag{6}$$

for all $s, t \in \mathbb{R}_+$.

This process has been used to model rainfall along the Nile River Valley. The long-range dependence of the rainfall was determined by Harold Hurst.

As is the case with Brownian motion it is sometimes useful to consider the formal time derivative of a fractional Brownian motion. Some particularly useful properties of a FBM are noted now. If $a > 0$, then the processes $(B(at), t \geq 0)$ and $\left(a^H B(t), t \geq 0\right)$ have the same probability law. This property is called a sell-similarity property. A long-range dependence for a FBM occurs if the Hurst parameter H is in the interval $\left(\frac{1}{2}, 1\right)$.

For the Hurst parameter $H \in (1/2, 1)$, a FBM has a long-range dependence property, which means that

$$\sum_{n=0}^{\infty} r(n) = \infty \tag{7}$$

where

$$r(n) = \mathbb{E}[B(1)(B(n+1) - B(n))] \tag{8}$$

This property provides an indication that the distant past has a nontrivial influence on the present behavior of the process.

Now the linear-quadratic control problem is reviewed [2–5]. Let $(X(t), t \geq 0)$ be the real-valued process that satisfies the stochastic differential equation

$$\begin{aligned} dX(t) &= \alpha_0 X(t)\,dt + bU(t)\,dt + dB(t) \\ X(t) &= X_0 \end{aligned} \tag{9}$$

where α_0, b, X_0 are constants, $(B(t), t \geq 0)$ is a standard fractional Brownian motion with the Hurst parameter $H \in (1/2, 1)$, $\alpha_0 \in [a_1, a_2]$, where $a_2 < 0$, $b \in \mathbb{R} \backslash \{0\}$.

Consider the optimal control problem where the state $X(t)$ satisfies Eq. (9) and the ergodic (or average cost per unit time) cost function J is

$$J(U) = \limsup_{T \to \infty} \frac{1}{T} \int_0^T \left(qX^2(t) + rU^2(t)\right) dt \tag{10}$$

where $q > 0$ and $r > 0$ are constants. The family $\mathcal{U}$ of admissible controls is all $(\mathcal{F}_t)$-adapted processes such that Eq. (9) has one and only one solution.

Let $H \in (1/2, 1)$ be fixed and $B(t)$ be a fractional Brownian motion with Hurst parameter H. For the applications in this paper, only a few results from a stochastic calculus are necessary. Let $f : [0, T] \to \mathbb{R}$ be a Borel measurable function. If f satisfies the condition:

$$|f|^2_{L^2_H} = \rho(H) \int_0^T \left(u_{1/2-H}(s) \left| I_{T-}^{H-1/2}\left(u_{H-1/2} f\right)(s) \right|\right)^2 ds < \infty \tag{11}$$

then $f \in L^2_H$ and $\int_0^T f\,dB$ is a zero mean Gaussian random variable with second moment

$$\mathbb{E}\left[\left(\int_0^T f\,dB\right)^2\right] = |f|^2_{L^2_H} \tag{12}$$

where $u_a(s) = s^a$ for $a > 0$ and $s \geq 0$, $I_{T-}^{H-1/2}$ is a fractional integral defined almost everywhere and given by

$$\left(I_{T-}^{H-1/2} f\right)(x) = \frac{1}{\Gamma(\alpha)} \int_x^T \frac{f(t)}{(t-x)^{3/2-H}}\,dt \tag{13}$$

for $x \in [0, T)$, $f \in L^1([0, T])$ and $\Gamma(\cdot)$ is the gamma function and ρ in Eq. (11) is

$$\rho(H) = \frac{H\Gamma(H + 1/2)\Gamma(3/2 - H)}{\Gamma(2 - 2H)} \tag{14}$$

If α_0 is unknown, then it is important to find a family of strongly consistent estimators of the unknown parameter α_0 in Eq. (9). A method is used that is called pseudo-least squares because it uses the least-squares estimation for α_0 assuming $H = \frac{1}{2}$, that is B, is then a standard Brownian motion. It can be shown that the family of estimators $(\hat{\alpha}(t)), t \geq 0)$ is strongly consistent $H \in (1/2, 1)$ for where

$$\hat{\alpha}(t) = \alpha_0 + \frac{\int_0^t X^0(s)dB(s)}{\int_0^t \left(X^0(s)\right)^2 ds} \tag{15}$$

and

$$dX^0(t) = \alpha_0 X^0(t)dt + dB(t) \tag{16}$$

$$X^0(0) = X_0 \tag{17}$$

An adaptive control $(U^\wedge(t), t \geq 0)$, is obtained from the certainty equivalence principle, that is, at time t, the estimate $\hat{\alpha}(t)$ is assumed to be the correct value for the parameter. Thus, the stochastic equation for the system (9) with the control $U^\wedge$ is

$$\begin{aligned} dX^\wedge(t) &= (\alpha_0 - \hat{\alpha}(t) - \delta(t))X^\wedge(t)dt \\ &\quad - \frac{b\rho(t)}{r} V^\wedge(t)dt + dB(t) \\ &= (-\alpha_0 - \hat{\alpha}(t) - \delta(t))X^\wedge dt \\ &\quad -(\alpha(t) + \delta(t))V^\wedge(t)dt + dB(t) \\ X^\wedge(0) &= X_0 \end{aligned} \tag{18}$$

where

$$\begin{aligned} \delta(t) &= \left[\alpha^2(t) + \frac{b^2}{r} q\right]^1 /2 \\ U^\wedge(t) &= \frac{b\rho(t)}{r}[X^\wedge(t) + V^\wedge(t)] \\ \rho(t) &= \frac{r}{b^2}[\alpha(t) + \delta(t)] \\ V^\wedge(t) &= \int_0^t \delta(s)V^\wedge(s)ds \\ &\quad + \int_0^t \left[\tilde{k}(t,s) - 1\right][dX^\wedge(s) - \alpha(s)X^\wedge(s)ds - bU^\wedge(s)ds] \\ &= \int_0^t \tilde{\delta}(s)V^\wedge(s)ds \\ &= \int_0^t \left[\tilde{k}(t,s) - 1\right][dB(s) + (\alpha_0 - \alpha(t))X^\wedge(s)ds] \\ &\textit{and} \\ \tilde{\delta}(t) &= \delta(t) + \alpha(t) - \alpha_0 \end{aligned} \tag{19}$$

The following result is obtained using the above [2].

Theorem 3.2. Let the scalar-valued control system satisfy the system equation given above. Let $(\alpha^\wedge(t), t \geq 0)$ be the family of estimators of α_0 given above, let $(U^\wedge(t)$ be the associated adaptive control given above and let $(X^\wedge(t)$ be the solution of the system with the control $U^\wedge$. Then the following limits exist:

$$lim_{t\to\infty} \frac{1}{t}\mathbf{E}\int_0^{\mathbf{t}} |\mathbf{U}^*(\mathbf{s}) - \mathbf{U}^\wedge(\mathbf{s})|^2 \mathbf{ds} = \mathbf{0} \tag{20}$$

$$lim_{t\to\infty} \frac{1}{t}\mathbf{E}\int_0^{\mathbf{t}} |\mathbf{X}^*(\mathbf{s}) - \mathbf{X}^\wedge(\mathbf{s})|^2 \mathbf{ds} = \mathbf{0} \tag{21}$$

so

$$lim_{t\to\infty} \frac{1}{t}\mathbf{E}\int_0^{\mathbf{t}} \left(\mathbf{q}(\mathbf{X}^\wedge(\mathbf{s}))^2 + \mathbf{r}(\mathbf{U}^\wedge(\mathbf{s}))^2\right)\mathbf{ds} = \lambda \tag{22}$$

where λ is the optimal cost for the known system.

A family of estimators can be obtained from Eq. (9) by removing the control term. The family of estimators $\alpha^\wedge$ is modified here using the fact that $\alpha_0 \in [a_1, a_2]$ as

$$\alpha^\wedge(t) = \hat{\alpha}(t)11_{[a_1,a_2]}(\hat{\alpha}(t)) \tag{23}$$

$$+a_1 11_{(-\infty,a_1)}(\hat{\alpha}(t)) + a_2 11_{(a_2,\infty)}(\hat{\alpha}(t)) \tag{24}$$

For $t \geq 0$. $\hat{\alpha}(0)$ is chosen arbitrarily in $[a_1, a_2]$. Then, it can be verified that a family of weighted least-squares estimates is strongly consistent for the unknown system parameter and that the family of running costs converges to the optimal cost. Thus, there is an optimal adaptive control for the system with a fractional Brownian motion. These results are summarized in the following theorem see T. E. Duncan et al. [2] for details and references.

Theorem 3.3. The family of estimators $(\alpha^\wedge(t), t \geq 0)$ is strongly consistent for the unknown system parameter and that the family of running costs converge to the optimal cost. Thus there is an optimal adaptive control for the partially known linear system with a fractional Brownian motion.

4. Noise modeled by a Rosenblatt process

Since there is significant evidence from physical systems (Domanski [13]) that Gaussian processes are inadequate for modeling noise in interconnected physical control systems, the family of Rosenblatt processes is considered. Some of the evidence is industrial processes and actuators are non-linear, they are highly interconnected and they have real external disturbances.

Initially, some definitions of a Rosenblatt process [14], a related fractional Brownian motion and some differential operators are given that fix notation and that are used in a change of variables formula and the solution of questions of absolute continuity.

Some fractional Brownian motions are naturally associated with a Rosenblatt process. Similar to fractional Brownian motion, Rosenblatt processes are indexed by the Hurst parameter $H \in (0, 1)$. Recall that both of these processes are members of the Hermite family of processes which are indexed by the number of multiple Wiener-Itô stochastic integrals. They are of order one and order two in the Hermite family reflecting the number of multiple Wiener-Itô integrals that are used to define them.

Let h_1^H and h_2^H be given as

$$h_1^H(u,y) = (u-y)^{H-\frac{3}{2}} \tag{25}$$

$$h_2^H(u,y_1,y_2) = (u-y_1)^{\frac{H}{2}-1}(u-y_2)^{\frac{H}{2}-1} \tag{26}$$

that are two singular kernels defined on the real line. These two kernels are used to define fractional Brownian motions and Rosenblatt processes. Rosenblatt processes like fractional Brownian motions are parametrized by the parameter H.

Definition 4.1. Let $H \in (1/2, 1)$ be fixed. A real-valued *fractional Brownian motion*, $B_H = (B_H(t), t \in R)$, is defined as follows

$$B_H(t) = C_H^B \int_R \left(\int_0^t h_1^H(u,y)du \right) dW(y) \tag{27}$$

For $t \geq 0$ (and similarly for $t < 0$), where C_H^B is a constant given below such that $\mathbb{E}(B_H^2(1)) = 1$ and W is a standard Wiener process (Brownian motion) on a fixed complete probability space denoted $(\Omega, \mathcal{F}, \mathbb{P})$ that is used throughout this paper.

Definition 4.2. Let $H \in (1/2, 1)$ be fixed. A real-valued (standard) *Rosenblatt process*, $R_H = (R_H(t), t \in \mathbb{R})$, is defined as follows

$$R_H(t) = C_H^R \int_{\mathbb{R}^2} \left(\int_0^t h_2^H(u,y_1,y_2)du \right) dW(y_1)dW(y_2) \tag{28}$$

For $t \geq 0$ (and similarly for $t < 0$), where C_H^R is a constant such that $\mathbb{E}(R_H^2(1)) = 1$, and the double stochastic integral is a Wiener-Itô multiple integral of order two with respect to the Wiener process (standard Brownian motion) W. This double Wiener integral is the definition given by Itô so the integral has expectation zero.

The normalizing constants C_H^B and C_H^R in the above two definitions are

$$C_H^B = \sqrt{\frac{H(2H-1)}{\beta(2-2H, H-\frac{1}{2})}}, \qquad C_H^R = \frac{\sqrt{2H(2H-1)}}{2\beta(1-H, \frac{H}{2})} \tag{29}$$

where β is the Beta function. For the subsequent change of variables (Itô-type) formula, it is also convenient to define the following constants

$$c_H^B = C_H^B \Gamma\left(H - \frac{1}{2}\right), \quad c_H^R = C_H^R \Gamma^2\left(\frac{H}{2}\right), \tag{30}$$

and

$$c_H^{B,R} = \frac{c_R^H}{c_{\frac{H}{2}+\frac{1}{2}}^B} = \sqrt{\frac{(2H-1)}{(H+1)} \frac{\Gamma(1-\frac{H}{2})\Gamma(\frac{H}{2})}{\Gamma(1-H)}} \tag{31}$$

where Γ is the Gamma function.

Some of the stochastic calculus for Rosenblatt processes is given in Refs. [11, 15]. A change of variables (Itô-Doeblin) formula is given in Čoupek et al. [11] that is used here to determine explicit Radon-Nikodym derivatives analogous to an approach for Wiener measure. The subsequent application of the change of variables formula for Rosenblatt processes uses the following two differential operators,

$$\nabla^{\frac{H}{2}} = I_{+}^{\frac{H}{2}} D \tag{32}$$

$$\nabla^{\frac{H}{2},\frac{H}{2}} = I_{+,+}^{\frac{H}{2},\frac{H}{2}} D^2. \tag{33}$$

where D is the Malliavin derivative and the other terms are the following fractional integrals

$$I_{+}^{\alpha}(f(x)) = \int_{-\infty}^{x} f(u)(x-u)^{\alpha-1} du \tag{34}$$

$$\begin{aligned}\left(I_{+,+}^{\alpha_1,\alpha_2} f\right)(x_1,x_2) = &\frac{1}{\Gamma(\alpha_1)\Gamma(\alpha_2)} \int_{-\infty}^{x_1}\int_{-\infty}^{x_2} f(u,v) \\ &(x_1-u)^{\alpha_1-1}(x_2-v)^{\alpha_2-1} du\, dv\end{aligned} \tag{35}$$

These operators reflect the singular integral definition of a Rosenblatt process. It follows from the direct use of the differential operator (33) that

$$\left(\nabla^{\frac{H}{2},\frac{H}{2}} R_H(t)\right)(u,u) = \tilde{C}_H \int_0^t |u-r|^{2H-2} dr \tag{36}$$

where the constant $\tilde{C}_H$, is given by

$$\begin{aligned}\tilde{C}_H &= 2c_H^R \frac{\beta^2\left(\frac{H}{2}, 1-H\right)}{\Gamma^2\left(\frac{H}{2}\right)} \\ &= \frac{1}{\Gamma\left(\frac{H}{2}\right)} \int_0^u (u-x)^{\frac{H}{2}-1} D_x W(s) dx\end{aligned} \tag{37}$$

and β is the Beta function.

The case where $t = u$ above is particularly useful here, so this case is explicitly given by

$$\begin{aligned}\left(\nabla^{\frac{H}{2},\frac{H}{2}} R_H(t)\right)(t,t) &= \tilde{C}_H \int_0^t |t-r|^{2H-2} dr = \tilde{C}_H \frac{t^{2H-1}}{2H-1} \\ &= \sqrt{\frac{2H}{2H-1}} \beta\left(\frac{H}{2}, 1-H\right) t^{2H-1}\end{aligned} \tag{38}$$

It is also necessary to note another derivative that appears in the change of variables formula associated with a Rosenblatt process.

$$\left(\nabla^{\frac{H}{2}}W(s)\right)(u) = \frac{1}{\Gamma\left(\frac{H}{2}\right)}\int_{-\infty}^{u}(u-x)^{\frac{H}{2}-1}D_x W(s)dx \tag{39}$$

$$= \frac{1}{\Gamma\left(\frac{H}{2}\right)}\int_{-\infty}^{u}1_{[0,s]}(x)(u-x)^{\frac{H}{2}-1}dx \tag{40}$$

$$= \frac{2}{H\Gamma\left(\frac{H}{2}\right)}\left[(u-s)_+^{\frac{H}{2}} - u_+^{\frac{H}{2}}\right] \tag{41}$$

where $W(t)$ is a standard Wiener process.

The change of variables results for stochastic equations driven by Rosenblatt processes is described now following [11]. Consider a scalar process $(y(t), t \geq 0)$ for the description of the change of variables result. Let $H \in \left(\frac{1}{2}, 1\right)$ be fixed, $y_0 \in \mathbb{R}$ be a constant, $\theta \in L^1_{loc}(0, \infty; D^{2,2})$, where $D^{2,2}$ is a Sobolev-Watanabe space, and define the process $(y(t), t \geq 0)$ by the equation

$$y(t) = y_0 + \int_0^t \theta(s)ds + R_H(t). \tag{42}$$

If some natural conditions [6] are satisfied for each $t > 0$, then the (scalar) process $(Y(t), t \geq 0)$ defined by $Y_t = f(t, y_t)$, where $f : \mathbb{R}_+ \times \mathbb{R} \to \mathbb{R}$ is a smooth function and satisfies the following equation

$$Y_t = Y_0 + \int_0^t \tilde{\theta}_s ds + 2c_H^{B,R}\int_0^t \tilde{\phi}_s dB_s^{\frac{H}{2}+\frac{1}{2}} + \int_0^t \tilde{\psi}_s dR_s^H \tag{43}$$

For each $t \geq 0$, (where)

$$\begin{aligned}
\tilde{\theta}_s &= \frac{\partial f}{\partial s}(s, y_s) + \frac{\partial f}{\partial x}(s, y_s)\theta_s \\
&\quad + c_H^R \frac{\partial^2 f}{\partial x^2}(s, y_s)\left(\nabla^{\frac{H}{2},\frac{H}{2}} y_s\right)(s, s) \\
&\quad + c_H^R \frac{\partial^3 f}{\partial x^3}(s, y_s)\left[\left(\nabla^{\frac{H}{2}} y_s\right)(s)\right]^2, \\
\tilde{\phi}_s &= \frac{\partial^2 f}{\partial x^2}(s, y_s)\left(\nabla^{\frac{H}{2}} y_s\right)(s), \\
\tilde{\psi}_s &= \frac{\partial f}{\partial x}(s, y_s).
\end{aligned} \tag{44}$$

It is useful to note that a third derivative occurs in the drift term and that a stochastic integral with respect to a fractional Brownian motion occurs.

As is the case for Brownian motion, this change of variables formula plays an important role in many stochastic calculations for a Rosenblatt process.

To indicate the ability to obtain explicit solutions for a Rosenblatt process, the explicit solution of an ergodic control problem for a scalar linear system is given, see P. Čoupek et al. [6] for details and references.

$$\begin{aligned} dX(t) &= aX(t)dt + bU(t)dt + dR^H(t) \\ X(0) &= x_0 \end{aligned} \tag{45}$$

The solution of this scalar stochastic system is

$$X_t = x_0 + \int_0^t (aX_s + bU_s)ds + R_t^H \tag{46}$$

where R^H is a Rosenblatt process with $H \in \left(\frac{1}{2}, 1\right)$ and $a, b, x_0 \in \mathbb{R}$, $b \neq 0$, are known constants.

The family of *admissible controls*, $\mathcal{U}$, is the collection of constant scalar linear feedback operators K, that is,

$$\mathcal{U} = \left\{(U_t)_{t \geq 0} : U_t = KX_t \,\text{with}\, K \in \mathbb{R}\right\}, \tag{47}$$

The following result is given in P. Čoupek et al. [6]:

Theorem. The optimal gain $\hat{K}$ in the family of admissible feedbacks is given by

$$\hat{K} = -\frac{a + \sqrt{a^2 + 4H(1-H)\frac{b^2 q}{r}}}{2b(1-H)}, \tag{48}$$

and the optimal cost $J_\infty(\hat{K})$ is given by

$$J_\infty(\hat{K}) = \frac{\Gamma(2H)}{\left[-(a + b\hat{K})\right]^{2H-1}} \left(\frac{-r\hat{K}}{b}\right). \tag{49}$$

The verification of this result for the optimal cost and an optimal control for a Rosenblatt noise is significantly more difficult than the corresponding results by replacing the Rosenblatt noise by a Brownian motion or a fractional Brownian motion, see Čoupek et al. [6] for details and many useful references.

5. Some future work

There are a number of natural directions to proceed for subsequent investigation. For multidimensional systems with either fractional Brownian motions or Rosenblatt processes, there is the basic problem of identification of unknown parameters. The basic parameters for estimation are those from the linear transformations A in the controlled equations for systems with either fractional Brownian motion or Rosenblatt noise or a combination of these noise processes. Furthermore, there are questions of optimal control for linear or nonlinear systems with fractional Brownian motions or Rosenblatt processes. Even an ergodic control problem for a scalar linear system with a Rosenblatt process with unknown system parameters needs to be addressed. Another family of models is those for infinite dimensional systems, especially for modeling some types of partial differential equations (PDEs). Many physical models are described by stochastic partial differential especially parabolic differential equations of second order. These equations often describe the physical situation of diffusion. Classically, a Brownian motion was developed to describe the diffusion of a liquid or gas. It seems quite likely that stochastic processes, such as fractional Brownian motions and Rosenblatt processes, will provide some of the mathematics for analysis of other physical phenomena.

6. Conclusions

This paper provides a comparison of results for adaptive control for linear systems driven by a Brownian motion, a fractional Brownian motion, and a Rosenblatt process. The results for a Brownian motion represent the initial noise process in a linear system and the results are fairly well developed. While Brownian motion for linear systems has been well developed over many years, the results for a fractional Brownian motion are significantly less developed. While these latter processes are Gaussian because they are Brownian motion processes, the fractional Brownian motion processes are not martingales and have a long-range dependence. These properties make it more difficult for analysis as contrasted with Brownian motion and their martingale properties. The Rosenblatt processes introduce more difficulties for analysis because they are neither martingales nor Gaussian processes. The development of useful tools for a Rosenblatt process is still in its infancy. Nevertheless, they are important and provide insight to the family of Hermite processes of which fractional Brownian motions are Hermite of order one and Rosenblatt processes are of order two.

Remark. For more background on Rosenblat‘’s processes, see Refs. [14, 15].

Acknowledgements

The authors thank F. Leve and F. Fahroo program directors at AFOSR for their support. Furthermore, the authors acknowledge many colleagues who have collaborated with the authors on identification and adaptive control, such as H. F. Chen, P. Coupek, L. Guo, Y. Hu, J. Jakubowski, B. Maslowski, L. Stettner, and many other colleagues and students.

Author details

Bozenna Pasik-Duncan* and Tyrone E. Duncan
University of Kansas, Lawrence, USA

*Address all correspondence to: bozenna@ku.edu

References

[1] Duncan TE, Guo L, Pasik-Duncan B. Adaptive continuous time linear quadratic Gaussian control. New York City: IEEE Transactions on Automatic Control. 1999;**44**:1653-1662

[2] Duncan TE, Pasik-Duncan B. Adaptive control of a scalar linear stochastic system with a fractional Brownian motion. In: Proceedings of the IFAC World Congress, Seoul. Oxford, UK: Elsevier Ltd.; 2008. pp. 4096-4101

[3] Duncan TE, Pasik-Duncan B. Linear quadratic fractional Gaussian control. SIAM Journal on Control and Optimization. 2013;**51**:4604-4619

[4] Duncan TE, Maslowski B, Pasik-Duncan B. Adaptive control for semilinear stochastic systems. SIAM Journal on Control and Optimization. 2000;**38**:1683-1706

[5] Kleptsyna ML, Le Breton A, Viot M. On the infinite-time horizon linear-quadratic regulatator problem under fractional Brownian perturbation. ESAIM Probability and Statistics. 2005;**9**: 185-204

[6] Čoupek P, Duncan TE, Maslowski B, Pasik-Duncan B. An infinite time horizon linear-quadratic control with a Rosenblatt process. In: Proceedings of the IEEE Conference on Decision and Control, Miami. Amsterdam: IOS Press; 2018

[7] Duncan TE, Hu YZ, Pasik-Duncan B. Stochastic calculus for fractional Brownian motion I. Theory. SIAM Journal on Control and Optimization. 2000;**38**:582-612

[8] Duncan TE, Maslowski B, Pasik-Duncan B. Semilinear stochastic equations in a Hilbert space with a fractional Brownian motion. SIAM Journal on Mathematical Analysis. 2009; **40**:2286-2315

[9] Duncan TE, Maslowski B, Pasik-Duncan B. Fractional Brownian motion and linear stochastic equations in a Hilbert space. Stochastics and Dynamics. 2002;**2**:225-250

[10] Duncan TE, Jakubowski J, Pasik-Duncan B. Stochastic integration for fractional Brownian motion in a Hilbert space. Stochastics and Dynamics. 2006;**5**: 53-75

[11] Čoupek P, Duncan TE, Pasik-Duncan B. A stochastic calculus for Rosenblatt processes. Stochastic Processes and their Applications. 2022. DOI: 10.1016/j.spa.2020.01.004

[12] Duncan TE. Some topics in fractional Brownian motion. Risk and Decision Analysis. 2009;**1**(3):145-153

[13] Domański PD. Non-Gaussian properties of the real industrial control error in SISO loops. In: 19th International Conference on System Theory, Control and Computing (ICSTCC). 2015. pp. 877-882

[14] Rosenblatt M. Independence and dependence. In: Proceedings of the Fourth Berkeley Symposium on Mathematical Statistics and Probability. Vol. 2. Berkeley, CA: University of California Press; 1961. pp. 431-443

[15] Tudor CA. Analysis of the Rosenblatt process. ESAIM: Probability and Statistics. 2008;**12**:230-257

Chapter 3

Information Recovery in Composite Model Reference Adaptive Control

Metehan Yayla and Ali Turker Kutay

Abstract

This study introduces a new adaptive control approach incorporating frequency-limited estimation of matched uncertainty. While many existing adaptive parameter adjustment laws aim to mitigate uncertainty effects solely through tracking error, it is well-documented that integrating uncertainty estimation error into the adaptation process significantly improves transient performance. Our method incorporates low-frequency uncertainty estimation with a time-varying learning rate structure. Unlike conventional filter-based approaches, our approach also compensates for information loss during signal filtering to suppress high-frequency content. Additionally, we include a regulation term in the standard adaptive weight update law, acting as stability enhancement in the adaptive system. We demonstrate the closed-loop stability of the proposed method using Lyapunov's stability theorem and highlight its efficacy through numerical examples and software-in-the-loop simulations with the X-plane flight simulator.

Keywords: filter, uncertainty, transients, fast adaptation, robustness

1. Introduction

Model Reference Adaptive Control (MRAC) frameworks achieve desired levels of closed-loop stability and performance for uncertain dynamical systems through online adaptive weight update laws. In the absence of restrictive persistent excitation (PE) of system signals, the standard MRAC framework cannot assure closed-loop stability when facing bounded perturbations using only instantaneous data [1]. To enhance the robustness of standard MRAC or ensure stability without PE, robust modifications are introduced in literature [2–7]. While these modifications guarantee the robustness of the adaptive controller, they generally yield limited enhancements in the transient response, primarily due to their focus on incorporating only the tracking error in the adaptation. Nonetheless, there exist a few additional modifications, such as those detailed in Refs. [8, 9], which notably improve transient performance by extracting additional information on uncertainty estimation error. Inspired by these studies, various modifications have been proposed in the literature, including but not limited to [10–16]. In these studies, information regarding either uncertainty estimation error or parameter convergence error is extracted using low-pass filters. However, filtering system signals may lead to information loss and degradation of adaptation performance.

To improve the tracking performance and achieve fast adaptation, it is desired to set the adaptive gains as high as possible. However, it is also well-practiced that the high learning rates can excite (possibly unmodeled) high-frequency modes of the system and result in system instability due to high-frequency oscillations. Numerous studies in the literature aim to facilitate the utilization of high gains while ensuring stability, such as the optimal control modification [5] and the low-frequency learning modification [17]. Yucelen and Haddad [17] introduced a modification term in the standard adaptive law to constrain the estimated parameters within bounds around their low-pass filtered signals. In the optimal control modification [5], high adaptive gains are permitted, albeit at the cost of reduced performance concerning robustness. Unlike these modifications, where the learning rate remains constant, there are studies where the adaptation gain varies over time [16, 18]. However, the learning rate vanishes in these methods if the basis function is not sufficiently and/or persistently exciting, causing the termination of adaptation.

So, our motivation becomes to combine these approaches in a way that the modification term allows the use of high adaptation gains without causing any high-frequency oscillations, and the learning rate of the proposed modification is adjusted to prevent undesired behavior during the adaptation.

In this study, we propose a new adaptive law that offers a modification term for information recovery in filter-based adaptive controllers. The proposed modification achieves fast adaptation by adding a new direction in adaptation using high-frequency content of the filtered signals. In addition, a time-varying learning rate is introduced to suppress the undesired effects of high-frequency signals on the system behavior. In this way, the proposed modification term behaves like a stability augmentation system to standard adaptive law. It should be noted that the proposed study builds upon the foundation established in the author's earlier work [19] by presenting more rigorous theoretical results and introducing additional simulations that highlight the effectiveness of the proposed method. Furthermore, the contributions are extended to cover a wider class of uncertain systems where limited information exists on uncertainty. These improvements result in a clearer and more precise presentation of the research, while expanding the domain of applications with new theoretical results through unstructured uncertainty parameterization.

2. Notation and definitions

The notation used in this study is fairly standard. Specifically, $\mathbb{R}$ denotes the set of real numbers, $\mathbb{R}^n$ denotes the set of $n \times 1$ real column vectors, $\mathbb{R}^{n\times m}$ denotes the set of $n \times m$ real matrices, $\mathbb{R}_+$ denotes the set of strictly positive real numbers, $(\cdot)^{-1}$ denotes inverse, $(\cdot)^T$ denotes transpose, '$\triangleq$' denotes equality by definition, $A>0$ denotes that A is a positive definite matrix, $(\cdot)^\dagger$ denotes the Moore–Penrose inverse, 'vec$(\cdot)$' denotes the column stacking operator, 'tr$(\cdot)$' denotes the trace operator, $\lambda_{\min}(A)$ ($\lambda_{\max}(A)$) returns the minimum eigenvalue (the maximum eigenvalue) of matrix A, $\sigma_{\max}(A)$ ($\sigma_{\min}(A)$) returns the maximum (minimum) singular value of matrix A, and $|a|$ denotes the absolute value of scalar $a \in \mathbb{R}$. Furthermore, for the vector $x \in \mathbb{R}^n$ and matrix $A \in \mathbb{R}^{m\times n}$, the Euclidean vector norm $\|x\|$, induced matrix norm $\|A\|$, and Frobenius matrix norm $\|A\|_F$ are defined as:

$$\|x\| \triangleq \sqrt{\sum_{i=1}^{n} x_i^2}, \qquad \|A\| \triangleq \sqrt{\lambda_{\max}(A^T A)} = \sigma_{\max}(A), \qquad \|A\|_F \triangleq \sqrt{\mathrm{tr}(A^T A)}. \tag{1}$$

Kronecker product is denoted by '$\otimes$'. For the vector $\theta \in \mathbb{R}^k$ and convex function $f : \mathbb{R}^k \to \mathbb{R}$, the gradient operator $\nabla f(\theta)$ is $\nabla f(\theta) = \left[\frac{\partial f(\theta)}{\partial \theta_1} \quad \cdots \quad \frac{\partial f(\theta)}{\partial \theta_k}\right]^T$. Lastly, '$\blacksquare$' indicates the completion of a mathematical proof.

Definition 1.1 (Persistent Excitation (PE)) *The signal $\omega(t) \in \mathbb{R}^n$ is said to be persistently exciting over an excitation period of τ if and only if there exist positive constants $\tau, \beta \in \mathbb{R}_+$ such that the following inequality holds:*

$$\int_{t-\tau}^{t} \omega(s)\omega^T(s)ds \succcurlyeq \beta I_{n\times n}, \qquad \forall t \in \mathbb{R}_+ \tag{2}$$

for all time $\forall t \in \mathbb{R}_+$ where $I_{n\times n}$ is the identity matrix of dimension n.

Definition 1.2 (Interval Excitation (IE)) *The signal $\omega(t) \in \mathbb{R}^n$ is said to be interval exciting at time $t = T_e$ over an excitation period of τ if there exist positive constants $\tau, \beta \in \mathbb{R}_+$ such that the following inequality holds:*

$$\int_{t-\tau}^{t} \omega(s)\omega^T(s)ds \succcurlyeq \beta I_{n\times n}, \qquad t = T_e \in \mathbb{R}_+ \tag{3}$$

Remark 1.1. *One can realize that the interval excitation (IE) is far less demanding condition than the persistent excitation (PE) as IE condition must be satisfied for all time $\forall t \in \mathbb{R}_+$ to satisfy the PE condition. That is, persistency of excitation is uniform in time, whereas the interval excitation just holds for a particular time interval.*

3. Problem definition

In this section, the control system description is provided first. Then, standard MRAC and composite/combined MRAC solutions are briefly reviewed for the completeness of the chapter.

3.1 System description

Consider the following nonlinear uncertain dynamical system:

$$\dot{x}(t) = Ax(t) + B[u(t) + \Delta(x)], \qquad x(t_0) = x_0 \tag{4}$$

where $x \in \mathcal{D}_x \subset \mathbb{R}^n$ is the state vector with $\mathcal{D}_x$ being a sufficiently large compact set, $u \in \mathcal{D}_u \subset \mathbb{R}^m$ is the m-dimensional control input that lies in the admissible control set $\mathcal{D}_u$, $\Delta : \mathbb{R}^n \to \mathbb{R}^m$ is the map representing the unknown uncertainty, $A \in \mathbb{R}^{n\times n}$ is known system matrix. Furthermore, input matrix $B \in \mathbb{R}^{n\times m}$ has full column rank and is assumed to be known (consistent with the literature; for example [6, 7, 14, 20, 21]), which is necessary to compute low-frequency content of the uncertainty using its pseudo-inverse (detailed later in Section 3.3). The fundamental assumptions are: (i) the pair (A, B) is controllable, (ii) full state measurement is available for feedback.

Assumption 1.1 (Structured Uncertainty). *The uncertainty $\Delta(x)$ can be expressed as linear combination of known basis functions $\phi : \mathcal{D}_x \to \mathbb{R}^s$ as follows:*

$$\Delta(x) = W^T\phi(x) \tag{5}$$

where $W \in \mathbb{R}^{s\times m}$ is the unknown constant weight matrix with $\|W\| \leq \overline{w}, \overline{w} > 0$, and known function $\phi(x)$ is Lipschitz continuous over $\mathcal{D}_x$.

Assumption 1.2 (Unstructured Uncertainty). *Assumption 1.1 can be relaxed using universal approximators such as Single Hidden Layer Neural Networks (SHL-NN) [22] or Radial Basis Functions (RBF) [23] as follows:*

$$\Delta(x) = W^T\phi(x) + \varepsilon(x) \tag{6}$$

where the residual $\varepsilon(x)$ with $\|\varepsilon(x)\| \leq \varepsilon_0, \forall x \in \mathcal{D}_x$.

Remark 1.2. *The basis function $\phi(x)$ in Assumption 1.2 can be chosen such that the residual $\varepsilon(x)$ can be made arbitrarily small on a compact set $\mathcal{D}_x$ with a positive scalar upper bound ε_0. Various methods exist for selecting such appropriate basis functions as universal approximators. One such method is to employ Radial Basis Functions (RBF) as suggested by the universal approximation theorem [23, 24] while minimizing the approximation error $\varepsilon(x)$. Due to its simplicity and extensive use in the literature (for example [25–27]), we parameterize the unstructured uncertainty in Eq. (6) using RBFs given by:*

$$\phi_i(x) = \begin{cases} 1 & , i = 1 \\ exp\left(-\dfrac{\|x(t) - \overline{c}_i\|^2}{2\overline{\mu}_i^2}\right) & , i = 2, \cdots, s \end{cases} \tag{7}$$

with $\overline{c}_i \in \mathbb{R}^n$ and $\overline{\mu}_i \in \mathbb{R}_+$ being the center and width of the ith node, respectively [23]. It should be noted that in the structured uncertainty case, the basis ϕ could be any known function and does not necessarily consist of Gaussian kernels.

The primary goal in model reference adaptive control (MRAC) is to design a control input to ensure that the system trajectories track those of the reference model given by:

$$\dot{x}_r(t) = A_r x_r(t) + B_r r(t), \qquad x_r(t_0) = x_{r0} \tag{8}$$

with $A_r \in \mathbb{R}^{n\times n}$ being Hurwitz reference system matrix and $B_r \in \mathbb{R}^{n\times k}$ being reference input matrix. Positive-definite matrix $P \in \mathbb{R}^{n\times n}$ satisfies the following Lyapunov equation

$$A_r^T P + PA_r = -Q, \qquad P = P^T > 0 \tag{9}$$

for any user-defined positive definite matrix $Q = Q^T > 0$. The reference state vector and reference command are denoted by $x_r \in \mathcal{D}_x$ and $r \in \mathbb{R}^k$, respectively.

Assumption 1.3 (Matching Conditions). *As a standard assumption in adaptive control [28–30], the reference model satisfies the matching conditions. That is, there exist matrices K_x and K_r such that the following conditions hold:*

$$\begin{aligned} A_r &= A - BK_x \\ B_r &= BK_r \end{aligned} \tag{10}$$

The nominal control input $u_n(t)$, consisting of feedback and feedforward components, is designed by setting $\Delta(x) = 0$ in the plant dynamics in Eq. (4) as follows:

$$u_n(t) = -K_x x(t) + K_r r(t) \tag{11}$$

The adaptive controller augmented control input is given by:

$$u(t) \quad = u_n(t) + u_a(t) = -K_x x(t) + K_r r(t) - \hat{W}^T(t)\phi(x) \tag{12}$$

with $\hat{W}(t)$ being the online estimation of unknown weight W.

3.2 Standard MRAC review

Let the tracking error be defined as $e(t) \triangleq x_r(t) - x(t)$. Combining the control in Eq. (12), matched uncertainty in Eq. (5), reference model in Eq. (8), and uncertain system dynamics in Eq. (4) with structured uncertainty parameterization yields the following tracking error dynamics:

$$\dot{e}(t) = A_r e(t) - B\tilde{W}^T(t)\phi(x) \tag{13}$$

with $\tilde{W} \triangleq W - \hat{W}$ being the weight estimation error. In the standard adaptive control formulation, the baseline weight update law is given by:

$$\text{Standard MRAC Law} : \dot{\hat{W}} = -\Gamma\phi e^T PB \tag{14}$$

with $\Gamma > 0$ being positive definite learning rate. The closed-loop stability of the standard adaptive control system can be shown using radially unbounded Lyapunov function $\mathrm{V}(e, \tilde{W}) = e^T Pe + \mathrm{tr}\left(\tilde{W}^T \Gamma^{-1} \tilde{W}\right) > 0$ with $\dot{\mathrm{V}}(e, \tilde{W}) = -e^T Qe \leq 0$. Note that $\mathrm{V}(0, 0) = 0$ and $\mathrm{V}(e, \tilde{W}) > 0$, for all $(e, \tilde{W}) \neq (0, 0)$, $\forall t \in \mathbb{R}_+$. Since $\mathrm{V}(e, \tilde{W})$ is lower-bounded by zero, and its derivative $\dot{\mathrm{V}}(e, \tilde{W}) \leq 0$ is less than or equal to zero, Lyapunov function $\mathrm{V}(e, \tilde{W})$ approaches to a finite limit as $t \to \infty$. Hence, the boundedness of tracking error $e(t)$ and weight estimation error $\tilde{W}(t)$ is guaranteed. Since $e(t)$ and $x_r(t)$ are bounded, system states $x(t)$ and basis vector-function $\phi(x)$ immediately become bounded. With bounded $e(t), x(t), \phi(x)$, and $\tilde{W}(t)$, time derivative of tracking error dynamics $\dot{e}(t)$ becomes bounded, which ensures the boundedness of $\ddot{\mathrm{V}}(e, \tilde{W}) = -2e^T Q\dot{e}$, $\forall t \in \mathbb{R}_+$. It then follows from Barbalat's Lemma [31] that $\lim_{t\to\infty} \dot{\mathrm{V}}(e(t), \tilde{W}(t)) = 0$ which implies the asymptotic stability of tracking error $e(t)$; that is, $\lim_{t\to\infty} e(t) = 0$. If the basis vector-function $\phi(x)$ is persistently exciting, then the estimated parameters converge to their ideal values; that is, $\hat{W}(t) \to W$ as $t \to \infty$. See Ref. [29] for details.

Remark 1.3. *If the uncertainty $\Delta(x)$ is characterized as unstructured uncertainty as in Assumption 1.2, the tracking error dynamics become:*

$$\dot{e}(t) = A_r e(t) - B\tilde{W}^T(t)\phi(x) - B\varepsilon(x) \tag{15}$$

In this case, the boundedness of the estimated parameters is not guaranteed by the standard MRAC without persistently exciting $\phi(x)$. Several robust modifications have been introduced in the literature to enhance the robustness, including but not limited to σ-mod

[2], e-mod [3], optimal control-based modification [5], Kalman filter-based modification [7], and q-modification [9]. Furthermore, it is common practice to include the projection operator [32] to bound the parameters within the prescribed convex set.

Remark 1.4. *Consider the adaptive law with σ-modification as a solution to the stability issue outlined in Remark 1.3:*

$$\dot{\hat{W}} = -\Gamma\left[\phi e^T PB + \sigma\hat{W}\right] \tag{16}$$

where $\sigma \in \mathbb{R}_+$ being constant scalar design parameter. It can be shown using Lyapunov function $V = \frac{1}{2}e^T Pe + \frac{1}{2}tr\left(\tilde{W}^T\Gamma^{-1}\tilde{W}\right)$ that the closed-loop system is stable in the sense of Lyapunov, even if $\varepsilon(x) \neq 0$ without requiring persistent excitation of the basis function ϕ. Specifically, the error signals e and $\tilde{W}$ are ensured to be inside the compact set D_η which is defined as $\mathcal{D}_\eta \triangleq \left\{\eta(t) : \|\eta\| < \sqrt{\frac{\mu_2}{\mu_1}}\right\}$ with μ_1, μ_2, η being as the followings:

$$\begin{aligned}
\mu_1 &\triangleq \min\left\{\left(\frac{1}{2}\lambda_{\min}(Q) - c_1\right), \sigma(1 - c_2)\right\} \\
\mu_2 &\triangleq \frac{\|PB\|^2\varepsilon_0^2}{4c_1} + \frac{\sigma\overline{w}}{4c_2} \\
c_1 &\in \left(0, \frac{1}{2}\lambda_{\min}(Q)\right), \quad c_2 \in (0, 1) \\
\eta &\triangleq \left[e^T, vec\left(\tilde{W}^T\right)\right]^T
\end{aligned} \tag{17}$$

3.3 Composite/combined MRAC review

Although high gains enable fast adaptation in standard MRAC, it may degrade the transient performance by inducing high-frequency oscillations in the system and even cause instabilities in the presence of large uncertainties. Improving the transient performance in MRAC has always been an attractive problem [5, 11, 17, 20, 21, 33, 34]. As seen from these studies and references therein, including the uncertainty estimation error in the adaptation enhances transient performance significantly. This fact was first utilized in [8], called '*Composite Model Reference Adaptive Control*'. Nearly at the same time, a quite similar approach was introduced [35] as '*Combined Model Reference Adaptive Control*'. Later in [20], these two approaches are generalized to cover the multi-input multi-output systems. These methods are referred in this study as composite/combined model reference adaptive control (CMRAC) to acknowledge both studies. Basically, CMRAC starts with filtering the system dynamics in Eq. (4) as:

$$\dot{x}_f = Ax_f + B\left[u_f + \Delta_f\right] \tag{18}$$

with subscript '*f*' indicating that the signal is filtered through the following low-pass filters:

$$\begin{aligned}
\dot{x}_f &= \omega_f\left(x - x_f\right), & x_f(t_0) &= x_0 \\
\dot{u}_f &= \omega_f\left(u - u_f\right), & u_f(t_0) &= u_0 \\
\dot{\phi}_f &= \omega_f\left(\phi - \phi_f\right), & \phi_f(t_0) &= \phi(x_0)
\end{aligned} \tag{19}$$

where $\omega_f \in \mathbb{R}_+$ is the cut-off frequency that determines the bandwidth of low-pass filters. Rearranging Eq. (18) yields:

$$\Delta_f = B^{\dagger}\left[\dot{x}_f - Ax_f\right] - u_f \tag{20}$$

Note that the signals x_f, u_f, ϕ_f, $\dot{x}_f$ and $\dot{\phi}_f$ are all accessible through Eq. (19). It should also be noted that Δ_f can be expressed as $\Delta_f = W^T\phi_f + \varepsilon_f$ from Eq. (5). Hence, we get

$$W^T\phi_f + \varepsilon_f = B^{\dagger}\left[\dot{x}_f - Ax_f\right] - u_f \tag{21}$$

Once the low-pass filters are applied to the system dynamics, the low-frequency content of uncertainty, Δ_f, becomes an available signal for control purposes, as all the signals on the right-hand side of Eq. (20) are available. Frequency-limited estimation of the uncertainty is defined as:

$$\hat{\Delta}_f = \hat{W}^T\phi_f \tag{22}$$

Next, one can define the low-frequency uncertainty estimation error as:

$$\tilde{\Delta}_f \triangleq \Delta_f - \hat{\Delta}_f \tag{23}$$

$$= B^{\dagger}\left[\dot{x}_f - Ax_f\right] - u_f - \hat{W}^T\phi_f \tag{24}$$

Then, the modified update law in CMRAC is given by

$$\text{CMRAC Law}: \dot{\hat{W}} = \Gamma\left[-\phi e^T PB + \gamma_c \phi_f \tilde{\Delta}_f^T\right] \tag{25}$$

where the error signal $\tilde{\Delta}_f$ is available for feedback in control through Eq. (24) and $\gamma_c \in \mathbb{R}_+$ is a positive scalar learning rate for CMRAC. It should be noted that CMRAC modification term is gradient descent-based minimizing solution of the following optimization problem:

$$\begin{aligned} &\min_{\hat{W}} \mathcal{J} \\ &\mathcal{J}(\hat{W}) = \frac{1}{2}\|\Delta_f - \hat{\Delta}_f\|^2 = \frac{1}{2}\|\tilde{\Delta}_f\|^2 \end{aligned} \tag{26}$$

where the gradient of the cost function in Eq. (26) is given by:

$$\nabla\mathcal{J}_{\hat{W}}(\hat{W}) = -\phi_f\left[\Delta_f(x) - \hat{W}^T(t)\phi_f(x)\right]^T = -\phi_f\tilde{\Delta}_f^T \tag{27}$$

Hence, the CMRAC weight update law can also be written as follows:

$$\text{CMRAC Law}: \dot{\hat{W}} = \Gamma\left[-\phi e^T PB - \gamma_c \nabla\mathcal{J}_{\hat{W}}(\hat{W})\right] \tag{28}$$

In this section, we revisit the CMRAC formulation assuming structured uncertainty definition as this is the case in contributing reference studies [8, 20, 36], which implies $\varepsilon(x) = \varepsilon_f(x,t) = 0$. Then, the asymptotic stability of the closed-loop system with CMRAC weight update law in Eq. (25) can be shown with radially unbounded Lyapunov function $V(e, \tilde{W}) = \frac{1}{2}e^T Pe + \frac{1}{2}\mathrm{tr}\left(\tilde{W}^T \Gamma^{-1} \tilde{W}\right)$. Its time-derivative along the system trajectories in Eq. (13) and Eq. (25) is given by $\dot{V}(e, \tilde{W}) = -\frac{1}{2}e^T Qe - \gamma_c \mathrm{tr}\left(\tilde{W}^T \phi_f \phi_f^T \tilde{W}\right)$ which is also equal to $\dot{V}(e, \tilde{W}) = -\frac{1}{2}e^T Qe - \gamma_c \|\tilde{\Delta}_f\|^2$. With this result, boundedness of the Lyapunov function V is guaranteed, which implies the boundedness of the state tracking error $e(t)$ and weight estimation error $\tilde{W}(t)$. Next, following the similar discussions made in MRAC, one can show the boundedness of $\ddot{V}(e, \tilde{W}) = -\frac{1}{2}e^T Q\dot{e} - \gamma_c \tilde{\Delta}_f^T \left(-\dot{\hat{W}}^T \phi_f + \tilde{W}^T \dot{\phi}_f\right)$ using Eqs. (19) and (25). It then follows from Barbalat's Lemma [31] that $\lim_{t\to\infty} \dot{V}(e(t), \tilde{W}(t)) = 0$ which implies the asymptotic stability of the state tracking error $e(t)$ and filtered uncertainty estimation error $\tilde{\Delta}_f(t)$; that is, $\lim_{t\to\infty} e(t) = 0$ and $\lim_{t\to\infty} \tilde{\Delta}_f(t) = 0$. It should be emphasized that the parameter convergence is still not guaranteed with CMRAC since the asymptotic stability of $\tilde{\Delta}_f$ does not imply the asymptotic stability of weight estimation error $\tilde{W}$. Furthermore, in case of unstructured uncertainty with nonzero residual $\varepsilon(x) \neq 0$, the boundedness of adaptive weights is not guaranteed without persistent excitation of $\phi(x)$ as in standard MRAC formulation, and the comments in Remark 1.3 apply to the CMRAC, as well.

Remark 1.5. *Cut-off frequency, ω_f, in Eq. (19) should be chosen carefully to reject the measurement noise and high-frequency content in the uncertainty. Although too small ω_f successfully filters the high-frequency signals in the adaptation, it may degrade the adaptation performance since $\hat{W}$ response becomes significantly sluggish [36].*

4. Main result

Although CMRAC and its further modifications (for example [10, 12–14, 21, 37]) contributed a lot to the robustness and performance of the standard MRAC, all these filtering-based solutions suffer from losing information during filtering. As emphasized in Remark 1.5, the performance degradation due to lost information may reach significant levels due to a poor selection of filter bandwidth. If the system experiences high-frequency variations (for example, payload drop for an aircraft, actuator failures), sluggish evolution of the adaptive weights may not guarantee the desired tracking performance. Furthermore, a relatively small cut-off frequency ω_f may delay the adaptation due to time-delay nature of low-pass filters. In this study, we address these issues and propose a new model reference adaptive control architecture that compensates for the information lost during filtering by including the high-frequency content of the filtered signals. In addition, a time-varying learning rate is also proposed to suppress the undesired effects of high-frequency signals in the adaptation.

Assumption 1.4 (Interval Excitation). *Basis $\phi \in \mathbb{R}^s$ is an interval exciting signal. Hence, there exist constant scalars $\tau, \alpha, T_e \in \mathbb{R}_+$ such that the following inequality holds:*

$$\int_{t-\tau}^{t} \phi(s)\phi^T(s)ds \succcurlyeq \alpha I_{s\times s}, \qquad t = T_e \tag{29}$$

Remark 1.6. *Assumption 1.4 ensures that the basis function ϕ contains as many spectral lines as the number of unknown parameters within time interval $t \in [T_e - \tau_w, T_e]$. Note that the basis function ϕ is an exogenous signal to the low-pass filter system in Eq. (19). Then, the filtered basis function ϕ_f also has the same number of spectral lines with less energy than that of the original basis ϕ. This implies the following:*

$$\int_{t-\tau}^{t} \phi_f(s)\phi_f^T(s)ds \geq \beta I_{n\times n}, \qquad t = T_e \tag{30}$$

with $0 < \beta \leq \alpha$. Let the transfer function representing the linear filter dynamics be $\mathcal{G}(s) \triangleq \frac{\omega_f}{s+\omega_f}$. Then, the degradation in the richness of the signal (which is equivalent to how small is β than α) depends on the cut-off frequency ω_f and spectrum of the basis ϕ. Let the spectral measure of ϕ is given by $S_\phi([\omega_1 \omega_2])$ with $\omega_1 < \omega_2$. Choosing the filter design parameter ω_f to be larger than ω_2 leads to relatively small degradation in the richness of ϕ_f with β being closer to α.

4.1 Integrated uncertainty estimation error

We further manipulate the low-frequency content of the matched uncertainty Δ_f by defining the following integrals over a moving window with a length of τ_ω seconds:

$$\begin{aligned} \mathcal{M}_\omega &\triangleq \int_{t-\tau_\omega}^{t} \phi_f \Delta_f^T d\xi = \left(\int_{t-\tau_\omega}^{t} \phi_f \phi_f^T d\xi\right) W + \int_{t-\tau_\omega}^{t} \phi_f \varepsilon_f^T d\xi \\ \hat{\mathcal{M}}_\omega &\triangleq \left(\int_{t-\tau_w}^{t} \phi_f \phi_f^T d\xi\right)\hat{\Theta} = \Phi_\omega \hat{\Theta} \end{aligned} \tag{31}$$

where $\hat{\Theta} \in \mathbb{R}^{s\times m}$ is an auxiliary weight, which will be discussed in Section 4.2.

Although ideal weight matrix W and filtered residual ε_f are unknown, the low-frequency content of the uncertainty, Δ_f, is accessible from Eq. (22). Hence, the integrated signal $\mathcal{M}_\omega$ can be computed during online operations. Next, the corresponding error $\tilde{\mathcal{M}}_\omega$ is defined as the following:

$$\tilde{\mathcal{M}}_\omega \triangleq \mathcal{M}_\omega - \hat{\mathcal{M}}_\omega = \left(\int_{t-\tau_\omega}^{t} \phi_f \phi_f^T d\xi\right)\tilde{\Theta} + \int_{t-\tau_\omega}^{t} \varepsilon_f \phi_f^T d\xi = \Phi_w \tilde{\Theta} + \int_{t-\tau_\omega}^{t} \varepsilon_f \phi_f^T d\xi \tag{32}$$

with $\tilde{\Theta} \triangleq W - \hat{\Theta}$ being an auxiliary weight estimation error. It should also be noted that the signal $\tilde{\mathcal{M}}_\omega$ is also available since both $\mathcal{M}_\omega$ and $\hat{\mathcal{M}}_\omega$ are accessible online. In order to preserve the information in excited basis functions, temporarily frozen variables $\mathcal{M}$ and $\hat{\mathcal{M}}$ are defined as follows:

$$\{\tilde{\mathcal{M}}(t), \Phi(t)\} \triangleq \begin{cases} \{\tilde{\mathcal{M}}_\omega(t), \Phi_\omega(t)\}, & \text{if } \kappa(\Phi_\omega(t)) < \kappa(\Phi(t^-)) \\ \{\tilde{\mathcal{M}}(t^-), \Phi(t^-)\}, & \text{otherwise} \end{cases} \tag{33}$$

where $\tilde{\mathcal{M}}(t^-)$ and $\phi(t^-)$ have zero initial conditions of appropriate dimensions. Additionally, $\kappa(\cdot)$ returns the condition number of its argument. With the definition in Eq. (33), it is ensured that the condition number of $\Phi(t)$ is monotonically non-increasing.

4.2 Auxiliary weight update law

In this section, we construct the first layer of dual adaptive law using the integrated signals in Section 4.1. The main purpose of the auxiliary adaptive law is to increase the stiffness of primary update law (which will be detailed in Section 4.3) and ensure the boundedness of adaptive weights $\hat{W}$ when $\Delta(x)$ is parameterized as unstructured uncertainty; i.e., $\varepsilon(x) \neq 0$. Additionally, the auxiliary adaptation guarantees the parameter convergence without PE if the uncertainty is structured and Assumption 1.4 on interval excitation holds; i.e., $\tilde{W} \to 0$ as $t \to \infty$ when $\varepsilon(x) = 0$. In order to construct the auxiliary weight update law, we begin with the following optimization problem:

$$\min_{\hat{W}} \mathcal{C}, \quad \mathcal{C}(\hat{W}) = \frac{1}{2}\|\mathcal{M} - \hat{\mathcal{M}}\|^2 \tag{34}$$

where the corresponding gradient becomes:

$$\nabla \mathcal{C}_{\hat{W}}(\hat{W}) = -\Phi\left(\mathcal{M} - \hat{\mathcal{M}}\right) = -\Phi(\Phi\tilde{\Theta} + \epsilon) \tag{35}$$

with $\epsilon = \mathcal{M} - \Phi W$. Then, the adaptive law for the auxiliary weights $\hat{\theta}$ is constructed using gradient descent-based optimization as:

$$\dot{\hat{\Theta}} = \Gamma_\Theta \Phi \tilde{\mathcal{M}}, \qquad \hat{\Theta}(t_0) = \hat{\Theta}_0 \tag{36}$$

with $\Gamma_\Theta \succ 0$ being positive definite gain matrix of appropriate dimensions.

Theorem 1.1. *Consider the uncertain dynamical system in Eq. (4) with unstructured uncertainty characterization in Eq. (6), auxiliary weight update law in Eq. (36), and integrated frozen signals in Eq. (33). Then, auxiliary weight estimation error $\tilde{\Theta}$ is uniformly ultimately bounded. Furthermore, if Assumption 1.4 on interval excitation of the basis function ϕ is satisfied, the estimated auxiliary weights converge to a closer neighborhood of ideal weight W.*

Proof: Consider the following Lyapunov function:

$$\mathrm{V}_\Theta = \frac{1}{2}\mathrm{tr}\left(\tilde{\Theta}^T \Gamma_\Theta^{-1} \tilde{\Theta}\right) \tag{37}$$

Its time-derivative along the trajectory in Eq. (36) is obtained as:

$$\dot{\mathrm{V}}_\Theta = -\mathrm{tr}\left(\tilde{\Theta}^T \Phi \tilde{\mathcal{M}}\right) = -\mathrm{tr}\left(\tilde{\Theta}^T \Phi\Phi\tilde{\Theta}\right) - \mathrm{tr}\left(\tilde{\Theta}^T \Phi\epsilon\right) = -\|\Phi\tilde{\Theta}\|_F^2 - \mathrm{tr}\left(\tilde{\Theta}^T \Phi\epsilon\right) \tag{38}$$

If Assumption 1.4 is not satisfied, $\Phi(t) = 0$ holds due to Eq. (33) with zero initial condition $\Phi(t_0) = 0$, which results in $\dot{\mathrm{V}}_\Theta = 0$. Thus, the adaptation for the auxiliary weights vanishes and $\hat{\Theta}$ stays at its initial condition, i.e. $\dot{\hat{\Theta}} = 0$, $\hat{\Theta}(t) = \hat{\Theta}_0$. Hence, the boundedness of auxiliary parameters is guaranteed if Assumption 1.4 does not hold. From this point forward, let us investigate the case where the interval excitation assumption holds. Note that, at any time t, the integrated residual error ϵ for the unstructured uncertainty case can be bounded from above as:

$$\begin{aligned}\left\|\int_{t-\tau_w}^{t} \varepsilon_f \phi_f^T d\xi\right\| &\leq \int_{t-\tau_w}^{t} \left\|\varepsilon_f \phi_f^T\right\| d\xi \leq \int_{t-\tau_w}^{t} \|\varepsilon_f\| \left\|\phi_f^T\right\| d\xi \\ &\leq \int_{t-\tau_w}^{t} \varepsilon_0 \sqrt{s} d\xi \leq \varepsilon_0 \tau_w \sqrt{s}, \qquad \forall t \in \mathbb{R}_+\end{aligned} \tag{39}$$

by utilizing the Gaussian kernels as radial basis functions for $\phi \in \mathbb{R}^s$. Then, the time-derivative of the Lyapunov function can be bounded as

$$\dot{V}_\Theta \leq -\|\Phi\tilde{\Theta}\|_F^2 + c_1\|\Phi\tilde{\theta}\|_F^2 + \frac{1}{4c_1}\|\epsilon\|^2 = -(1-c_1)\|\Phi\tilde{\Theta}\|_F^2 + c_2^2 \tag{40}$$

with $c_2 \triangleq \frac{\varepsilon_0 \tau_w \sqrt{s}}{2\sqrt{c_1}}$. Then, its is clear that for any $c_1 \in (0, 1)$, the matrix product $\Phi\tilde{\Theta}$ is bounded since $\dot{V}_\Theta < 0$ if $\|\Phi\tilde{\Theta}\|_F > c_2\sqrt{1/(1-c_1)}$. As Assumption 1.4 is satisfied, the matrix Φ is positive definite with $\lambda_{\min}(\Phi) \geq \beta$ from Remark 1.6. The boundedness of $\Phi\tilde{\Theta}$ with an invertible Φ guarantees the boundedness of auxiliary parameter estimation error $\tilde{\Theta}$, since $\tilde{\Theta} = \Phi^{-1}(\Phi\tilde{\Theta})$, that is, $\|\tilde{\Theta}\| \leq d_1$ with $d_1 > 0$ being unknown positive constant that depends on the excitation level β, residual bound ε_0, and number of RBF neurons s. With constant unknown weight matrix W and $\tilde{\Theta} \in \mathcal{L}_\infty$, the boundedness of auxiliary weight estimation $\hat{\Theta}$ is ensured, as well. ■

Corollary 1.1. *Consider the control system outlined in Theorem 1.1. With structured uncertainty parameterization in Eq. (5), auxiliary weights $\hat{\Theta}$ converge exponentially to the ideal weights W if Assumption 1.4 holds.*

Proof: Structured uncertainty parameterization implies $\varepsilon(x) = \epsilon = 0$. Hence, Lyapunov derivative becomes:

$$\dot{V}_\Theta = -\|\Phi\tilde{\Theta}\|_F^2 \leq -\lambda_{\min}(\Phi^2)\|\tilde{\Theta}\|^2 \leq -\beta^2\|\tilde{\Theta}\|^2 < 0 \tag{41}$$

with β being excitation level in Remark 1.6. Quadratic positive definite Lyapunov function with quadratic negative definite derivative ensures the exponential convergence of the auxiliary weight convergence error $\tilde{\Theta}$ at a rate of β^2; that is, $\|\tilde{\Theta}(t)\| \leq \|\tilde{\Theta}(t_0)\|e^{-\beta^2(t-t_0)} \triangleq d_2(t) \leq \overline{d}_2$ with $\overline{d}_2 \in \mathbb{R}_+$ being an unknown constant scalar. Since the Lyapunov function is radially unbounded, this result holds globally.

Remark 1.7. *Unlike the minimum singular value maximization algorithm [38], the condition number minimization is utilized in Eq. (33). The main reason is that the deviation from the ideal solution W in least-squares regression due to perturbation ε is directly proportional to the condition number of the regressor matrix Φ. Therefore, minimizing the condition number of Φ improves the robustness against residual error $\varepsilon(x)$. Furthermore, maximizing the minimum singular value is inherently embedded in condition number minimization due to the definition $\kappa(\cdot) = \frac{\lambda \max(\cdot)}{\lambda \min(\cdot)}$. It should be emphasized that the condition number minimization also introduces additional information of $\lambda_{\max}(\Phi)$ on temporarily frozen matrices, which results in more frequent update, more robust adaptation, and faster development of weight estimations. Further discussion on condition number minimizing algorithm can be found in [39].*

4.3 Primary layer of adaptation

In this section, the primary weight update law is introduced for the dual adaptive controller. We start by decomposing the weight update law as:

$$\dot{\hat{W}} = \dot{\hat{W}}_b + \dot{\hat{W}}_m, \qquad \text{with} \quad \dot{\hat{W}}_b = -\Gamma\phi e^T PB \tag{42}$$

where $\dot{\hat{W}}_b$ is the standard MRAC law in Eq. (14) and $\dot{\hat{W}}_m$ is the proposed modification term given as the following:

$$\dot{\hat{W}}_m = \left(\Gamma^{-1} + \gamma_m\phi_f\phi_f^T\right)^{-1}\left[\gamma_m\phi_f\phi_f^T\Gamma\phi e^T PB + \gamma_m\dot{\phi}_f\tilde{\Delta}_f^T + \gamma_1\phi_f\tilde{\Delta}_f^T + \gamma_2(\hat{\Theta} - \hat{W})\right] \tag{43}$$

where $\gamma_1, \gamma_2, \gamma_m \in \mathbb{R}_+$ are user-defined design variables. Then, the proposed weight update law can be written as

$$\dot{\hat{W}} = -\mathbb{X}\phi e^T PB + \mathbb{X}\left[\gamma_1\phi_f\tilde{\Delta}_f^T + \gamma_2(\hat{\Theta} - \hat{W}) + \gamma_m\dot{\phi}_f\tilde{\Delta}_f^T\right] \tag{44}$$

where the time-varying learning rate $\mathbb{X}$ is given by

$$\mathbb{X} \triangleq \left(\Gamma^{-1} + \gamma_m\phi_f\phi_f^T\right)^{-1} \tag{45}$$

Lemma 1.1. *The Learning rate* $\mathbb{X}$ *is a positive definite matrix.*
Proof:

$$\begin{aligned} \mathbb{X} \succ 0 \qquad & \Rightarrow \left(\Gamma^{-1} + \gamma_m\phi_f\phi_f^T\right)^{-1} \succ 0 \\ & \Rightarrow \Gamma^{-1} + \gamma_m\phi_f\phi_f^T \succ 0 \\ \left(\text{since}\ '\phi_f\phi_f^T \succcurlyeq 0'\ \text{and}\ \gamma_m > 0\right) & \Rightarrow \Gamma^{-1} \succ 0 \quad \Rightarrow \quad \Gamma \succ 0 \end{aligned} \tag{46}$$

Since $\Gamma \succ 0$ by the design, the claim $\mathbb{X} \succ 0$ holds, as well.

Remark 1.8. *The term* $\gamma_2(\hat{\Theta} - \hat{W})$ *is essentially a σ-modification term introduced to the adaptive law, causing the estimated weight* $\hat{W}$ *to remain bounded around auxiliary weight* $\hat{\Theta}$*. Recall that* $\hat{\Theta}$ *is bounded from Theorem 1.1. Hence, the stability results from standard MRAC with σ-modification in Remark 1.4 can directly be applied to ensure the Lyapunov stability of all the closed-loop signals with the proposed adaptive controller, as will be shown by Theorem 1.2 and Theorem 1.3.*

4.4 Stability results for unstructured uncertainty

Theorem 1.2. *Consider the uncertain system dynamics given in Eq. (4), uncertainty parametrization in Assumption 1.2, reference model in Eq. (8), control input in Eq. (12), filter states in Eq. (19), and adaptive law in Eq. (44). Then, all the signals in closed-loop system are uniformly ultimately bounded.*

Proof: Consider the following radially unbounded Lyapunov function:

$$\mathrm{V} = \frac{1}{2}e^T Pe + \frac{1}{2}\mathrm{tr}\left(\tilde{W}^T\mathbb{X}^{-1}\tilde{W}\right) \le \frac{1}{2}\lambda_{\max}(P)\|e\|^2 + \frac{1}{2}\left(\lambda_{\max}(\Gamma^{-1}) + \gamma_m s\right)\|\tilde{W}\|^2 \tag{47}$$

Note that $\mathrm{V}(0,0) = 0$ and $\mathrm{V}(e,\tilde{W}) \neq 0$ for all $(e,\tilde{W}) \neq (0,0)$, $\forall t \in \mathbb{R}_+$. Time derivative of the Lyapunov function along the system trajectories in Eqs. (13) and (44) is given as follows:

$$\begin{aligned}\dot{V} &= e^T P\dot{e} + \operatorname{tr}\left(\tilde{W}^T \mathbb{X}^{-1}\dot{\tilde{W}}\right) + \frac{1}{2}\operatorname{tr}\left(\tilde{W}^T \mathbb{X}^{-1}\tilde{W}\right) \\ &= -\frac{1}{2}e^T Qe - \operatorname{tr}\left(\tilde{W}^T \mathbb{X}^{-1}\dot{\hat{W}}\right) + \gamma_m \operatorname{tr}\left(\tilde{W}^T \dot{\phi}_f \phi_f^T \tilde{W}\right)\end{aligned} \tag{48}$$

Substituting the weight update law in Eq. (44) into Eq. (48) gives:

$$\begin{aligned}\dot{V} = &-\frac{1}{2}e^T Qe - \gamma_1 \operatorname{tr}\left(\tilde{W}^T \phi_f \phi_f^T \tilde{W}\right) - \gamma_2 \operatorname{tr}\left(\tilde{W}^T \tilde{W}\right) \\ &-\gamma_1 \operatorname{tr}\left(\tilde{W}^T \phi_f \varepsilon_f^T\right) - \gamma_m \operatorname{tr}\left(\tilde{W}^T \dot{\phi}_f \varepsilon_f^T\right) - \gamma_2 \operatorname{tr}\left[\tilde{W}^T (\hat{\Theta} - W)\right] \\ = &-\frac{1}{2}e^T Qe - \gamma_1 \|\tilde{W}^T \phi_f\|_F^2 - \gamma_2 \|\tilde{W}\|_F^2 \\ &-\gamma_1 \operatorname{tr}\left(\tilde{W}^T \phi_f \varepsilon_f^T\right) - \gamma_m \operatorname{tr}\left(\tilde{W}^T \dot{\phi}_f \varepsilon_f^T\right) - \gamma_2 \operatorname{tr}\left[\tilde{W}^T (\hat{\Theta} - W)\right] \\ \leq &-\frac{1}{2}\lambda_{\min}(Q)\|e\|^2 - \gamma_1 \|\tilde{W}^T \phi_f\|^2 - \gamma_2 \|\tilde{W}\|^2 + \|\tilde{W}\| d_4 \\ = &-\frac{1}{2}\lambda_{\min}(Q)\|e\|^2 - \gamma_1 \|\tilde{W}^T \phi_f\|^2 - \gamma_2 \|\tilde{W}\|\left(\|\tilde{W}\| - d_4\right)\end{aligned} \tag{49}$$

where $d_4 \triangleq \gamma_1 \varepsilon_0 \sqrt{s} + \gamma_m \varepsilon_0 d_3 + \gamma_2 d_1$, $\|\hat{\Theta} - W\| \leq d_1$ is guaranteed from Theorem 1.1, $\|\dot{\phi}_f\| \leq d_3$ holds for $\phi(x)$ consisting of smooth Gaussian kernels, with $d_3 > 0$ being an unknown positive constant. It is obvious that the time derivative of Lyapunov function is negative definite for $\|\tilde{W}\| > d_4$. Thus, the error signals e and $\tilde{W}$ asymptotically converge to a compact set, ensuring $e, \tilde{W} \in \mathcal{L}_\infty$. Since the reference model state x_r is bounded and the unknown weight matrix W is constant, boundedness of system state vector x and adaptive weights $\hat{W}$ is guaranteed.

4.5 Stability results for structured uncertainty

Theorem 1.3. *Consider the uncertain system dynamics given in Eq. (4), uncertainty parametrization in Assumption 1.1, reference model in Eq. (8), control input in Eq. (12), filter states in Eq. (19), and adaptive law in Eq. (44). Then, the state tracking error $e(t)$ is asymptotically stable. Furthermore, all the signals in a closed-loop system are bounded. If the basis function ϕ is interval exciting, then the zero-solution $(e, \tilde{W}, \tilde{\Delta}_f) = (0,0,0)$ is asymptotically stable.*

Proof: Recall the Lyapunov function in Theorem 1.2:

$$V = \frac{1}{2}e^T Pe + \frac{1}{2}\operatorname{tr}\left(\tilde{W}^T \mathbb{X}^{-1}\tilde{W}\right) \leq \frac{1}{2}\lambda_{\max}(P)\|e\|^2 + \frac{1}{2}\left(\frac{1}{\lambda_{\min}(\Gamma)} + \gamma_m s\right)\|\tilde{W}\|^2 \tag{50}$$

with its time derivative being

$$\dot{V} = -\frac{1}{2}\lambda_{\min}(Q)\|e\|^2 - \gamma_1 \|\tilde{W}^T \phi_f\|^2 - \gamma_2 \|\tilde{W}\|\left(\|\tilde{W}\| - d_4\right) \tag{51}$$

where $d_4 = \gamma_1 \varepsilon_0 \sqrt{s} + \gamma_m \varepsilon_0 d_3 + \gamma_2 d_1$. With structured uncertainty parameterization, we have $\varepsilon = \varepsilon_f = \epsilon = 0$, yielding $d_4 = \gamma_2 d_1$ and $\tilde{\Delta}_f = \tilde{W}^T \phi_f$. The discussions for boundedness of all system signals in Theorem 1.2 also hold in structured uncertainty cases with a smaller bound on $\tilde{W}$. It can also be shown that the error signals $e, \tilde{\Delta}_f, \tilde{W}$

are asymptotically stable by incorporating the results from Corollary 1.1. Specifically, upper bound d_1 can be replaced with $d_2(t)$, i.e., $d_4 = \gamma_2 d_2(t)$ where $d_2(t) \to 0$ as $t \to \infty$ exponentially at a rate of β^2. Thus,

$$\dot{V}(t) \to -\frac{1}{2}\lambda_{\min}(Q)\|e\|^2 - \gamma_1\|\tilde{\Delta}_f\|^2 - \gamma_2\|\tilde{W}\|^2 < 0, \qquad \text{as } t \to \infty \tag{52}$$

Therefore, the tracking error e low-frequency uncertainty estimation error $\tilde{\Delta}_f$, and weight estimation error $\tilde{W}$ asymptotically converge to zero, ensuring the asymptotic stability of the zero-solution $\left(e, \tilde{W}, \tilde{\Delta}_f\right) = (0{,}0{,}0)$.

4.6 Discussion

The proposed information recovery-based composite model reference adaptive controller (IR-CMRAC) acquires all the benefits of CMRAC as IR-CMRAC update law already contains the minimizing solution in Eq. (27). Additionally, the proposed method exhibits significant improvements as the following:

- As the design parameter γ_m tends to zero, the learning rate $\mathbb{X}$ in Eq. (45) tend to Γ, furthermore, with $\gamma_2 = 0$, the entire proposed modification term results in CMRAC law in Eq. (25). Hence, the proposed law can be considered as a generalized form of CMRAC.

- Time derivative of gradient of the cost function in Eq. (27) is given by

$$\frac{d}{dt}\left(\nabla \mathcal{J}_{\hat{W}}\right) = 2\dot{\phi}_f \phi_f^T \tilde{W} - \phi_f \phi_f^T \dot{\hat{W}}_b - \phi_f \phi_f^T \dot{\hat{W}}_m \tag{53}$$

  It can be observed that the proposed update law in Eq. (44) contains the derivative information given in Eq. (53). Incorporating the derivative into traditional gradient descent-based optimization allows the shape of the transient behavior. In that respect, it can be considered analogous to the derivative action in traditional proportional/integral/derivative (PID) controllers.

- Each term in Eq. (53) contributes to the adaptation performance in different manners. Specifically, the first term, '$\dot{\phi}_f \tilde{\Delta}_f$', compensates for the information lost while filtering the basis function ϕ by including the high-frequency content of basis ϕ. Recall that $\dot{\phi}_f$ is given by $\dot{\phi}_f = \omega_f\left(\phi - \phi_f\right)$; that is, $\dot{\phi}_f$ actually contains only the high-frequency components of basis ϕ. It should also be noted that the product of $\dot{\phi}_f$ and ϕ_f is still a high-frequency signal.

- The second term in Eq. (53), '$\phi_f \phi_f^T \dot{\hat{W}}_b$', regulates the standard MRAC through gain γ_m. In this context, this term can be considered analogous to proportional feedback control for the standard adaptive law, acting akin to a stability augmentation system. Essentially, it entails readjusting the learning rate Γ to prevent adverse effects on adaptation from large values in the basis function ϕ.

- The last term in Eq. (53), '$\phi_f \phi_f^T \dot{\hat{W}}_m$', regulates the learning rate of the modification term in a similar way to the standard adaptive law regulation explained in previous item.

- Last but not the least, term '$\gamma_2(\hat{\Theta} - \hat{W})$' in Eq. (44) plays a vital role in ensuring the parameter convergence in structured uncertainty parameterization; i.e. $\hat{W}(t) \to W$ as $t \to \infty$. In that respect, this term acts like an integral action in a traditional PID controllers as it analogously allows to remove the steady state error in the weight estimations. Specifically, the auxiliary weight $\hat{\Theta}$ evolves in the direction of minimizing the steady-state error in the parameter estimation, and the term '$\gamma_2(\hat{\Theta} - \hat{W})$' ensures that the adaptive weights $\hat{W}$ stay bounded around $\hat{\Theta}$, which inherently results in zero steady-state error in $\tilde{W}$. For unstructured uncertainty parameterization, boundedness of adaptive weights around close neighborhood of the ideal parameters are achieved (instead of parameter convergence).

4.7 Inverse-free weight update law

One might observe that the updated law proposed in Eq. (44) necessitates computing the inverse of a (potentially large matrix) during online operations. This would undoubtedly incur additional computational costs and practical challenges. To address these drawbacks, we proceed to manipulate this term to achieve an inverse-free learning rate.

Lemma 1.2. *Given that $\Gamma = \Gamma^T > 0$ is positive definite matrix, $\gamma_m \in \mathbb{R}_+$ is a positive constant scalar, and $\phi_f \in \mathbb{R}^s$, the matrix $\left(\Gamma^{-1} + \gamma_m \phi_f \phi_f^T\right)$ is invertible and its inverse is given by*

$$\left(\Gamma^{-1} + \gamma_m \phi_f \phi_f^T\right)^{-1} = \Gamma - \frac{1}{1 + tr\left(\gamma_m \phi_f \phi_f^T \Gamma\right)} \Gamma \gamma_m \phi_f \phi_f^T \Gamma > 0 \tag{54}$$

Proof: Note that $\Gamma = \Gamma^T > 0$ and $\gamma_m \phi_f \phi_f^T \geqslant 0$. This fact implies $\left(\Gamma^{-1} + \gamma_m \phi_f \phi_f^T\right)$ is also a symmetric positive definite matrix, i.e. $\left(\Gamma^{-1} + \gamma_m \phi_f \phi_f^T\right) > 0$. Then, $\left(\Gamma^{-1} + \gamma_m \phi_f \phi_f^T\right)$ is always invertible. Let us define the following for clarity:

$$\begin{aligned} E &\triangleq \gamma_m \phi_f \phi_f^T = \left(\phi_f \sqrt{\gamma_m}\right)\left(\phi_f \sqrt{\gamma_m}\right)^T \\ G &\triangleq \Gamma^{-1} \quad \to G^{-1} = \Gamma \end{aligned} \tag{55}$$

Then,

$$\begin{aligned} EG^{-1}E &= \left(\phi_f \sqrt{\gamma_m}\right)\left(\phi_f \sqrt{\gamma_m}\right)^T \Gamma \left(\phi_f \sqrt{\gamma_m}\right)\left(\phi_f \sqrt{\gamma_m}\right)^T \\ &= \left[\left(\phi_f \sqrt{\gamma_m}\right)^T \Gamma \left(\phi_f \sqrt{\gamma_m}\right)\right]\left(\phi_f \sqrt{\gamma_m}\right)\left(\phi_f \sqrt{\gamma_m}\right)^T \\ &= \mathrm{tr}\left[\left(\phi_f \sqrt{\gamma_m}\right)\left(\phi_f \sqrt{\gamma_m}\right)^T \Gamma\right]\left(\phi_f \sqrt{\gamma_m}\right)\left(\phi_f \sqrt{\gamma_m}\right)^T \\ &= \mathrm{tr}(EG^{-1})E \end{aligned} \tag{56}$$

Le $(G+E)^{-1}$ be in the form of $(G+E)^{-1} = \left(G^{-1} - \nu G^{-1}EG^{-1}\right)$. Then, their product

$$(G+E)\left(G^{-1} - \nu G^{-1}EG^{-1}\right) = I - \nu EG^{-1} + EG^{-1} - \nu EG^{-1}EG^{-1} \tag{57}$$

gives identity matrix if the equality $\left(\nu EG^{-1} - EG^{-1} + \nu EG^{-1}EG^{-1}\right) = 0$ is satisfied. Then, combining this inequality with the result in Eq. (56) yields:

$$\nu EG^{-1} - EG^{-1} + \nu tr\left(EG^{-1}\right)EG^{-1} = \left[\nu - 1 + \nu tr\left(EG^{-1}\right)\right]EG^{-1} = 0 \tag{58}$$

This equality holds for any square matrix E with rank$(E) = 1$ and invertible matrix G if the parameter v is chosen as

$$\nu = \frac{1}{1 + \mathrm{tr}\left(EG^{-1}\right)} = \frac{1}{1 + \mathrm{tr}\left(\gamma_m \phi_f \phi_f^T \Gamma\right)} \tag{59}$$

Substituting v into $(G+E)^{-1} = \left(G^{-1} - \nu G^{-1}EG^{-1}\right)$ results in

$$\left(\Gamma^{-1} + \gamma_m \phi_f \phi_f^T\right)^{-1} = \Gamma - \frac{1}{1 + \mathrm{tr}\left(\gamma_m \phi_f \phi_f^T \Gamma\right)} \gamma_m \Gamma \phi_f \phi_f^T \Gamma \tag{60}$$

That completes the proof. Readers may refer to [40] for details.

5. Illustrative numerical examples

5.1 Systems with integrator dynamics

In this example, the first order roll dynamics of an aircraft is considered:

$$\dot{p}(t) = L_p p(t) + L_\delta[\delta_a(t) + \Delta(p)] \tag{61}$$

where $p(t)$ is roll rate, $\delta_a(t)$ is aileron input, dynamic stability derivative L_p denoting the roll damping is $L_p = -1$, control derivative is $L_\delta = 1$, and the uncertainty $\Delta(p)$ is given by

$$\Delta(p) = 0.2 + p|p| \tag{62}$$

This numerical example can be regarded as an incorrect characterization of the roll mode time constant and the existence of constant wind disturbance trying to roll the aircraft. To improve the tracking performance robustness, an integrator state x_i is augmented to the system as follows:

$$\begin{bmatrix} \dot{p}(t) \\ \dot{x}_i(t) \end{bmatrix} = \begin{bmatrix} L_p & 0 \\ -1 & 0 \end{bmatrix} \begin{bmatrix} p(t) \\ x_i(t) \end{bmatrix} + \begin{bmatrix} L_\delta \\ 0 \end{bmatrix} [\delta_a(t) + \Delta(p)] + \begin{bmatrix} 0 \\ 1 \end{bmatrix} p_c(t) \tag{63}$$

where p_c is the commanded roll rate.

It is important to note that standard MRAC suffers from degraded performance in systems with augmented integrator dynamics if the learning rate is relatively high. As clearly seen in **Figure 1**, increasing the learning rate induces high-frequency oscillations in the system. In this example, the efficacy of the proposed method will also be highlighted for the systems with augmented integrator dynamics.

The reference model is chosen as:

$$\dot{x}_r(t) = \begin{bmatrix} -2.5 & 3 \\ -1 & 0 \end{bmatrix} x_r(t) + \begin{bmatrix} 0 \\ 1 \end{bmatrix} p_c(t) \tag{64}$$

Thus, the nominal controller becomes:

$$u_n(t) = -1.5p(t) + 3x_i(t) \tag{65}$$

5.1.1 Without integral action in the adaptation ($\gamma_2 = 0$)

Numerical simulations are carried out with sampling frequency of 100 Hz, positive definite solution P is determined using $Q = I_{2\times2}$, and low-pass filter bandwidth is $\omega_f = 4$ rad/s. In addition, adaptive gains are $\gamma_1 = \gamma_c = \gamma_m = 1$, $\Gamma = I_{2\times2}$, $\Gamma_\Theta = 2I_{2\times2}$, $\tau_w = 20$ s, and $\gamma_2 = 0$. That is, the integral action in adaptation is not activated for results shown in **Figures 2** and **3**. It should be noted that the common adaptive gains are kept the same to make a fair comparison; that is, $\gamma_1 = \gamma_c$ and Γ is constant for all simulations. Lastly, the commanded roll rate p_c is a square wave with an amplitude of 1 rad/s and period of 10 s (see **Figure 1**).

In **Figure 2**, state tracking performance for different adaptive controllers is illustrated. As clearly seen, the nominal controller results in oscillatory roll response due to insufficient damping. Among all the adaptive controllers, the best tracking performance is achieved by the IR-CMRAC with almost zero state tracking error. Additionally, the desired tracking response is realized by the aileron control input with no oscillations.

Figure 3 indicates the evolution of the weight estimation error $\tilde{W}$. Although the integral action is disabled with $\gamma_2 = 0$, the parameter convergence is achieved within 15 seconds. It should be noted that the exciting signal $p_c(t)$ is persistently exciting; hence, the parameter convergence is an expected result. However, the important point is that the fastest convergence is achieved with IR-CMRAC as more information

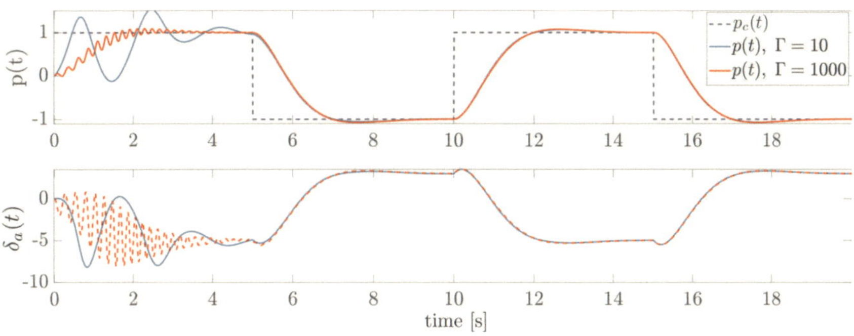

Figure 1.
Performance comparison for standard MRAC in the presence of high gains.

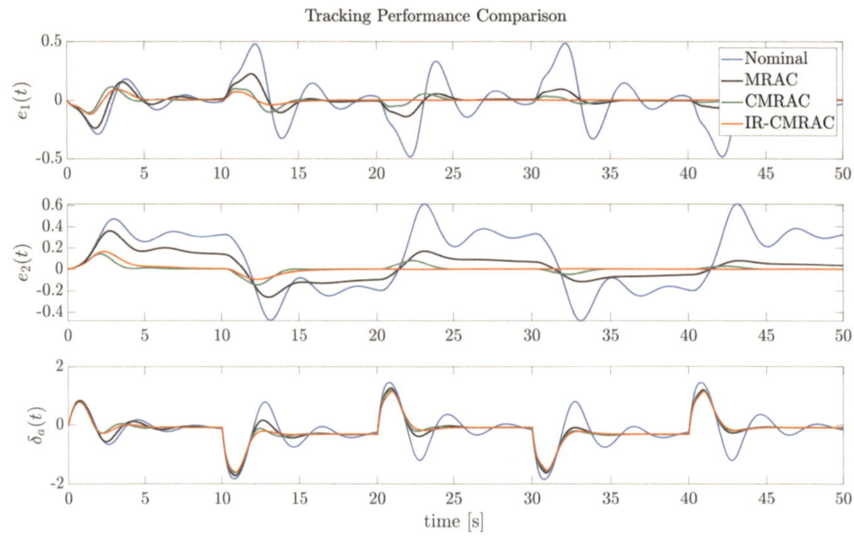

Figure 2.
Tracking performance comparison of adaptive controllers ($\gamma_2 = 0$ in IR-CMRAC).

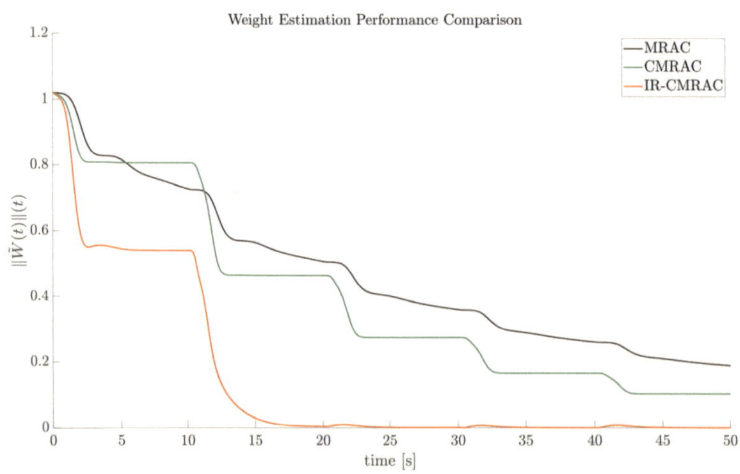

Figure 3.
Weight estimation performance comparison of adaptive controllers ($\gamma_2 = 0$ in IR-CMRAC).

is included in the adaptation, which adds new directions to update the adaptive weights.

Figure 4 illustrates the evolution of time-varying learning rate $\mathbb{X}$. One can observe that the variation of diagonal elements $\mathbb{X}_{11}$ and $\mathbb{X}_{22}$ is more apparent during the transients. This is because the stability augmentation system becomes active especially during the high-frequency variations in the signals. In this way, the undesired effects of high-frequency signals used in the adaptation are suppressed.

5.1.2 With integral action in the adaptation ($\gamma_2 = 10$)

In this scenario, the integral action in IR-CMRAC is activated by setting the learning rate $\gamma_2 = 10$ in Eq. (44). The state tracking performance, evolution of time-

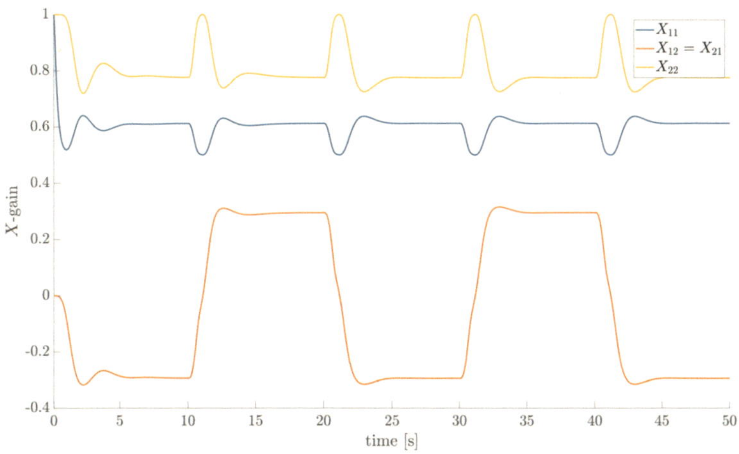

Figure 4.
Evolution of time-varying learning rates in IR-CMRAC ($\gamma_2 = 0$ in IR-CMRAC).

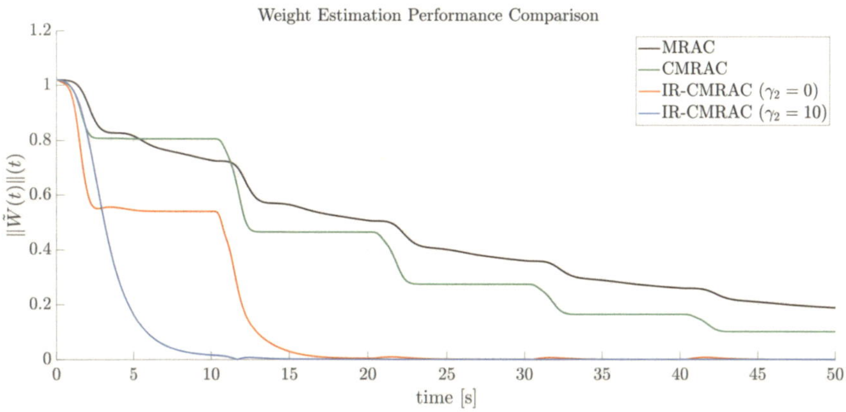

Figure 5.
Weight estimation performance comparison of adaptive controllers.

varying learning rate $\mathbb{X}$, and aileron control input are not illustrated for this case since the differences between simulation results with $\gamma_2 = 0$ and $\gamma_2 = 10$ are indistinguishable. However, the weight estimation performance is worth mentioning and presented in **Figure 5**. Unlike IR-CMRAC with $\gamma_2 = 0$, the adaptive weights keep evolving with $\gamma_2 = 10$ even if the systems states are not excited (for example on the interval of $t \in [5, 10]$). This is because, by the time $t = 5$ seconds, the basis function ϕ has become sufficiently exciting, thereby fulfilling Assumption 1.4. As a result, faster convergence of the adaptive weights is achieved, as claimed by Theorem 1.3.

5.2 Presence of measurement noise and external disturbance

In this example, an external disturbance is added to the uncertainty as:

$$\Delta(p, t) = 0.2 + 0.23e^{-0.13t} + 0.1\sin(0.2t) + 0.13\sin(0.37t) + p|p| \tag{66}$$

Newly introduced sinusoidal signals can be considered as the effects of variable wind speed and direction, whereas the exponential term is used to simulate the wind shear. With this definition of aggregated uncertainty, the unknown weight matrix becomes time-varying:

$$W(t) = \begin{bmatrix} 0.2 - 0.23e^{-0.13t} + 0.1\sin(0.2t) + 0.13\sin(0.37t) \\ 1 \end{bmatrix}, \quad \phi(p) = \begin{bmatrix} 1 \\ p|p| \end{bmatrix} \tag{67}$$

Furthermore, the measurement noise is added to the roll rate p as

$$p(t) \leftarrow p(t) + \upsilon(t) \tag{68}$$

where $\upsilon(t)$ is the Gaussian white noise with standard deviation of $2\,deg/s$, which practically ensures $\upsilon(t) \in [-6, 6]deg/s$, $\forall t \geq t_0$. All the simulation parameters are the same as in Section 5.1.1, except for the learning rate $\gamma_2 = 1$.

Figure 6 illustrates the state tracking performance of the proposed adaptive controller for the outlined simulation scenario. Due to the low-pass filter nature of integration, the integrator state x_2 is not affected drastically by the roll rate measurement noise. Furthermore, IR-CMRAC is mainly driven by the low-pass filtered signals (for example ϕ_f, $\tilde{\Delta}_f$, u_f in Eq. (44)); the effects of measurement noise are suppressed successfully, thereby resulting in smooth aileron control input. This result can also be observed in the time-varying learning rate components $\mathbb{X}_{ij}$ given in **Figure 7**.

It is important to remember that the convergence results of Theorem 1.3 are valid if the unknown weight matrix W is constant. However, it can readily be shown that all the system signals are uniformly ultimately bounded if the unknown weight matrix is bounded and time-varying, i.e. $\|W(t)\|_2 \leq \delta_w$ with $\delta_w \in \mathbb{R}_+$ being unknown upper bound. In **Figure 8**, the evolution of the adaptive weights is illustrated in the presence

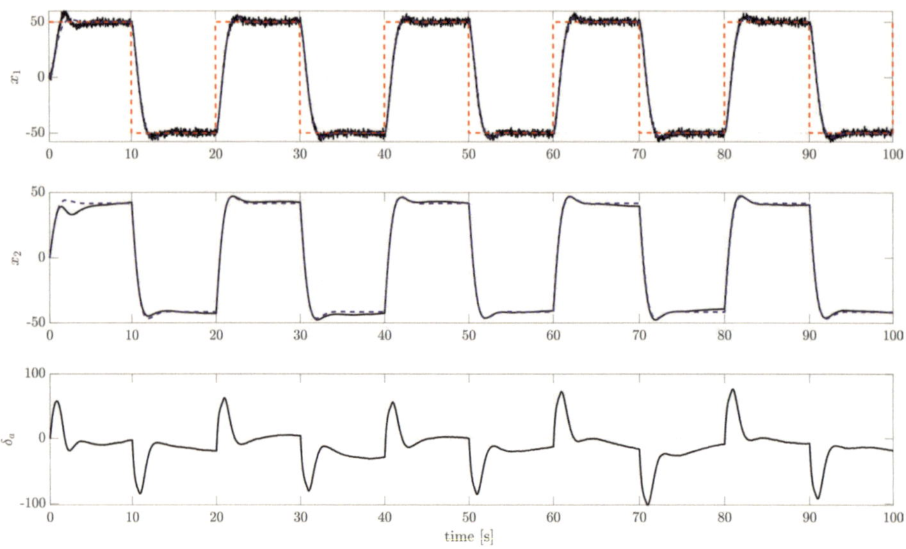

Figure 6.
State tracking performance for IR-CMRAC in the presence of noise and disturbance.

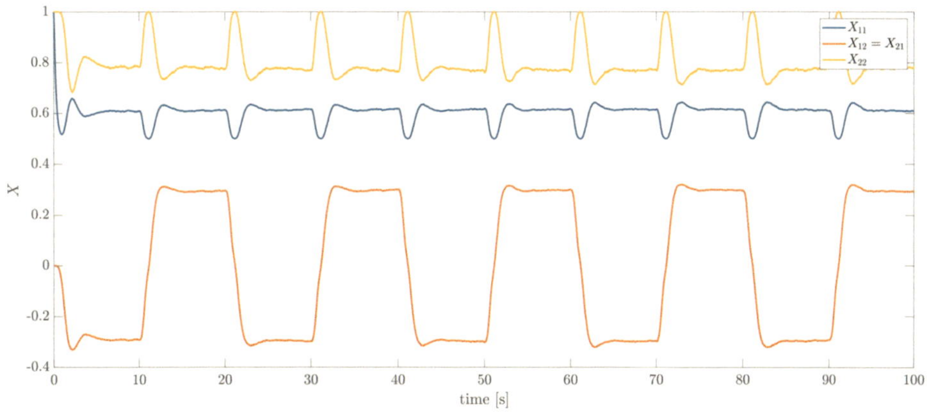

Figure 7.
Variation of components of the learning rate $\mathbb{X}$ for IR-CMRAC.

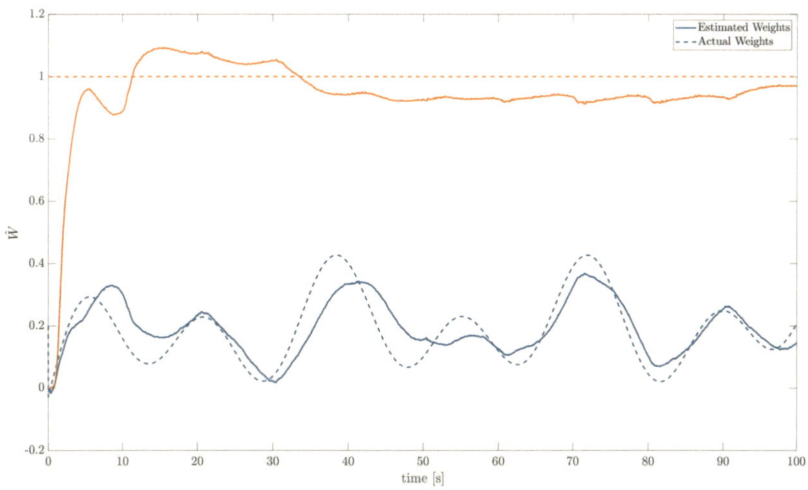

Figure 8.
Evolution of the adaptive weights for IR-CMRAC in the presence of measurement noise and disturbance (dashed line: Actual weights $W(t)$, solid line: Estimated weights $\hat{W}(t)$.)

of external disturbance and measurement noise. As seen in the figure, the adaptive parameters stay bounded around their actual values and are practically noise-free.

5.3 Investigation of information recovery

In this example, we conduct numerical simulations to highlight the main contribution of the proposed method: information recovery. In order to make a fair comparison with CMRAC, the integral action in the adaptation is deactivated with $\gamma_2 = 0$. All the other simulation parameters are as in Section 5.1.1. Measurement noise and disturbance are not applied to the system dynamics. Eventually, the following update laws are compared for the plant dynamics in Eq. (61) with the uncertainty in Eq. (62):

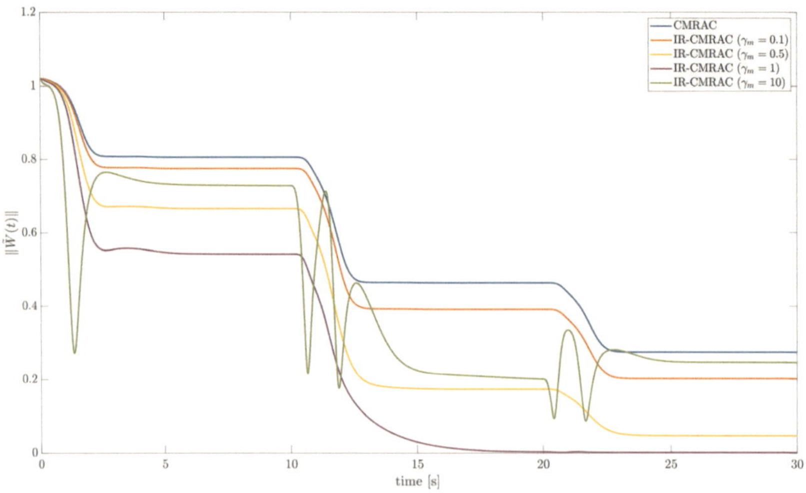

Figure 9.
The effect of γ_m on the weight estimation error for IR-CMRAC.

$$\begin{aligned}
&\text{CMRAC Law} &&: \dot{\hat{W}} = -\Gamma\phi e^T PB + \Gamma\gamma_1\phi_f\tilde{\Delta}_f^T \\
&\text{IR-CMRAC Law}(\gamma_2 = 0) &&: \dot{\hat{W}} = -\mathbb{X}\phi e^T PB + \mathbb{X}\left(\gamma_1\phi_f\tilde{\Delta}_f^T + \gamma_m\dot{\phi}_f\tilde{\Delta}_f^T\right)
\end{aligned} \tag{69}$$

Simulation results for several values of γ_m are shown in **Figure 9**. The lost information in CMRAC (due to low-pass filtering) is included in the adaptation with nonzero γ_m. Hence, the weight estimation error becomes smaller with increasing γ_m up to 1. However, further increase in γ_m results in overshoot and oscillatory response, causing degraded weight estimation performance. This can be considered analogous to the excessive derivative gain in a traditional *PID* controller, where the unnecessarily large derivative gains might cause instability. It is clear from the figure that judicious tuning of the gain γ_m results in significantly improved adaptive weight estimation and hence, the desired state tracking performance.

5.4 Adaptation with non-PE signals

The parameter convergence in both standard MRAC and CMRAC is dependent on the stringent requirement of persistent excitation of basis function $\phi(x)$. Note that PE condition of $\phi(x)$ depends on the persistent excitation of system states x [41]. Furthermore, states x are persistently exciting provided that the exogenous reference input $r(t)$ meets PE condition [4]. In this section, we provide simulation results with non-PE reference signal to illustrate the efficacy of the proposed method under interval excitation (IE) condition, which is much weaker than PE as described in Remark 1.1.

Figure 10 compares the state-tracking performances of MRAC, CMRAC, and IR-CMRAC. In the simulated flight case, the aircraft performs a right bank maneuver and returns back to wings-level flight condition exponentially.

As seen clearly from **Figure 10**, the reference signal is not persistently exciting but satisfies the interval excitation condition. The parameter convergence is achieved without persistent excitation, as suggested by Corollary 1.1 and Theorem 1.3 (see **Figure 11**).

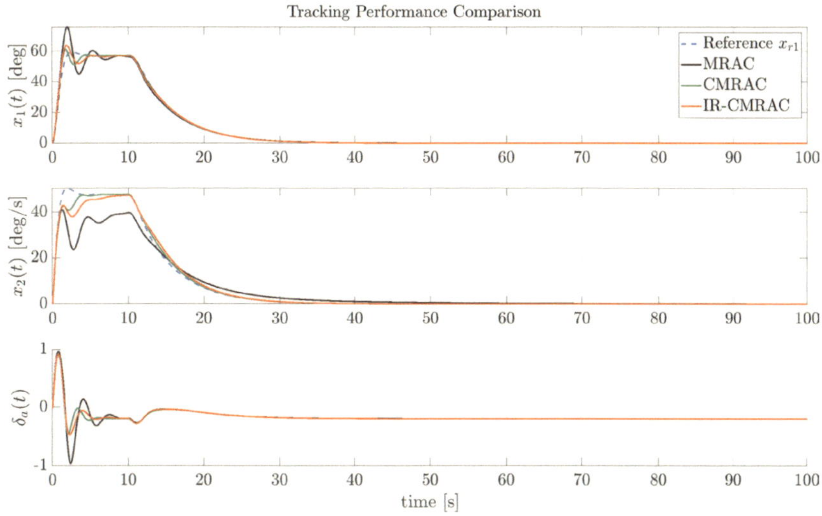

Figure 10.
State tracking performance with non-PE reference signal.

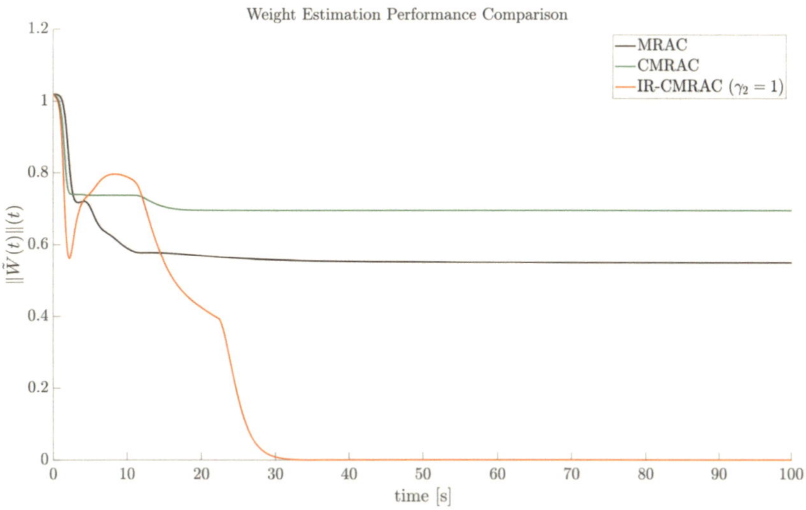

Figure 11.
Weight estimation error with non-PE reference signal.

6. Software-in-the-loop (SITL) simulations for lateral flight control

6.1 X-plane flight simulator and lateral dynamics

X-plane flight simulator operates according to the blade element theory, where the physics engine calculates velocity components for each blade element. These components stem from various sources like rotational motion, free-stream velocity, propeller inflow, and downwash and wake caused by aerodynamic surfaces and fuselage. Thus, the aircraft model realized by the X-plane flight simulator is highly nonlinear and realistic, which makes X-plane a Federal Aviation Administration (FAA) certified flight simulator [42].

Linear equations of motion for the lateral dynamics of F-4 Phantom II at steady wings-level trim conditions with true airspeed of $V_\infty = 360$ knots and altitude of $h =$ 24000 ft. are obtained using '*Athena Vortex Lattice*' [43] as:

$$\begin{bmatrix} \dot{\beta} \\ \dot{p} \\ \dot{r} \\ \dot{\phi} \end{bmatrix} = \underbrace{\begin{bmatrix} -1.47 & 0.0 & -1.0 & 0.075 \\ -18.85 & -7.52 & 2.89 & 0.0 \\ 15.15 & 0.56 & -7.61 & 0.0 \\ 0.0 & 1.0 & 0.0 & 0.0 \end{bmatrix}}_{\triangleq A_p} \underbrace{\begin{bmatrix} \beta \\ p \\ r \\ \phi \end{bmatrix}}_{\triangleq x_p} + \underbrace{\begin{bmatrix} 0.0 & -0.15 \\ 8.48 & -0.96 \\ -0.18 & 3.94 \\ 0.0 & 0.0 \end{bmatrix}}_{\triangleq B_p} \underbrace{\begin{bmatrix} \delta_a \\ \delta_r \end{bmatrix}}_{\triangleq u} \tag{70}$$

$$y_p = \begin{bmatrix} \beta \\ \phi \end{bmatrix} = \begin{bmatrix} 1 & 0 & 0 & 0 \\ 0 & 0 & 0 & 1 \end{bmatrix} x_p = E_p x_p$$

where β is sideslip angle $[rad]$, p and r are roll and yaw rates $[rad/s]$ (resp.), and ϕ is bank angle $[rad]$. Aileron and rudder inputs are normalized with maximum deflection of 20 degrees to be $\delta_a \in [-1, 1]$ and $\delta_r \in [-1, 1]$, respectively.

6.2 Nominal controller design

In order to increase the robustness and improve the tracking performance, integrator states are introduced as follows:

$$\begin{aligned} \dot{z}(t) &= \begin{bmatrix} \beta_{cmd}(t) \\ \phi_{cmd}(t) \end{bmatrix} - \begin{bmatrix} \beta(t) \\ \phi(t) \end{bmatrix} \\ \dot{z}(t) &= r(t) - y_p(t) = r(t) - E_p x_p(t) \end{aligned} \tag{71}$$

Then, augmented system dynamics can be expressed as follows:

$$\begin{bmatrix} \dot{x}_p \\ \dot{z} \end{bmatrix} = \underbrace{\begin{bmatrix} A_p & 0 \\ -E_p & 0 \end{bmatrix}}_{\triangleq A} \underbrace{\begin{bmatrix} x_p \\ z \end{bmatrix}}_{\triangleq x} + \underbrace{\begin{bmatrix} B_p \\ 0 \end{bmatrix}}_{\triangleq B} \begin{bmatrix} \delta_a \\ \delta_r \end{bmatrix} + \underbrace{\begin{bmatrix} 0 \\ I \end{bmatrix}}_{\triangleq B_r} \begin{bmatrix} \beta_{cmd} \\ \phi_{cmd} \end{bmatrix} \tag{72}$$

$$\dot{x} = Ax + Bu + B_r r$$

Reference model satisfying $A_r = A - BK$ is given as

$$\dot{x}_r = A_r x_r + B_r r \tag{73}$$

The nominal controller $u_n(t) = -Kx$ is designed using feedback gain K:

$$K = \begin{bmatrix} 1.14 & 1.15 & -0.81 & 4.14 & -3.25 & -3.02 \\ -0.58 & -0.69 & 0.09 & -1.32 & 4.76 & 0.97 \end{bmatrix} \tag{74}$$

The fastest mode is the roll subsidence with frequency of 7.66 rad/s. Hence, the cut-off frequency of the low-pass filter is chosen to be $\omega_f = 8$ rad/s.

6.3 SITL simulation results

In this section, we demonstrate SITL simulation results of the F-4 Phantom fighter jet's lateral dynamics. The objective is to execute consecutive left and right bank maneuvers (see **Figure 12**) while ensuring minimal sideslip to uphold coordinated flight. Initially, adaptation remains inactive for 70 seconds, and then the proposed adaptive controller is activated. The communication between the controller and the X-plane flight simulator is established through the user datagram protocol (UDP) port. Hence, an unknown communication delay is present in the simulations.

In **Figure 13**, state tracking performance is illustrated. As the proposed adaptation is activated, the sideslip angle gets smaller, resulting in better-coordinated bank maneuvers. Furthermore, the oscillations, especially in the roll channel, are significantly reduced with the adaptation.

Figures 14 and **15** illustrate the commanded and actuated control surface commands. It is important to highlight that the improvements in state tracking performance are not because of the excessive use of the controls but due to the effective

Figure 12.
Left and right Bank Maneuvers as the flight scenario.

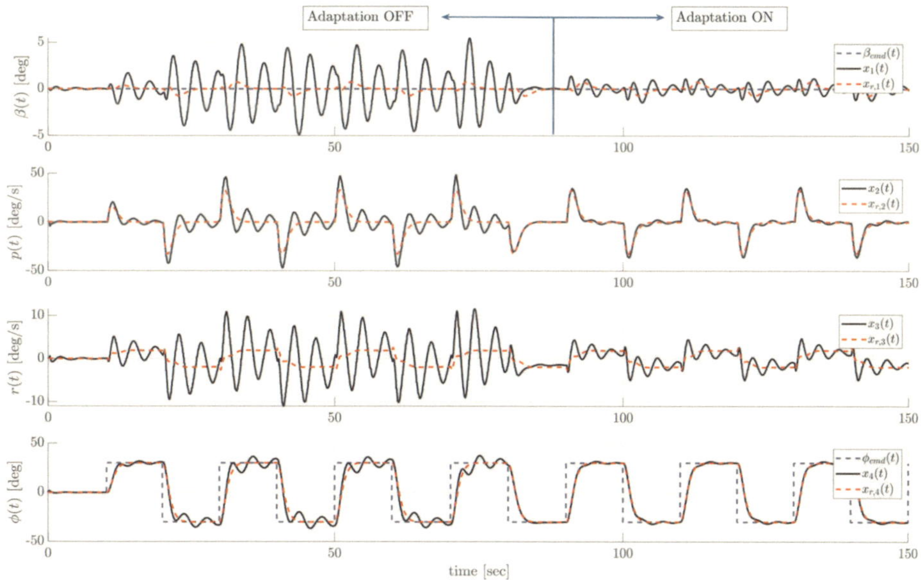

Figure 13.
State tracking performance for SITL simulation with $\Gamma = \Gamma_\Theta = 2I$, $\gamma_m = 10$, $\tau_w = 30$, $\gamma_1 = 1$, *and* $\gamma_2 = 10$.

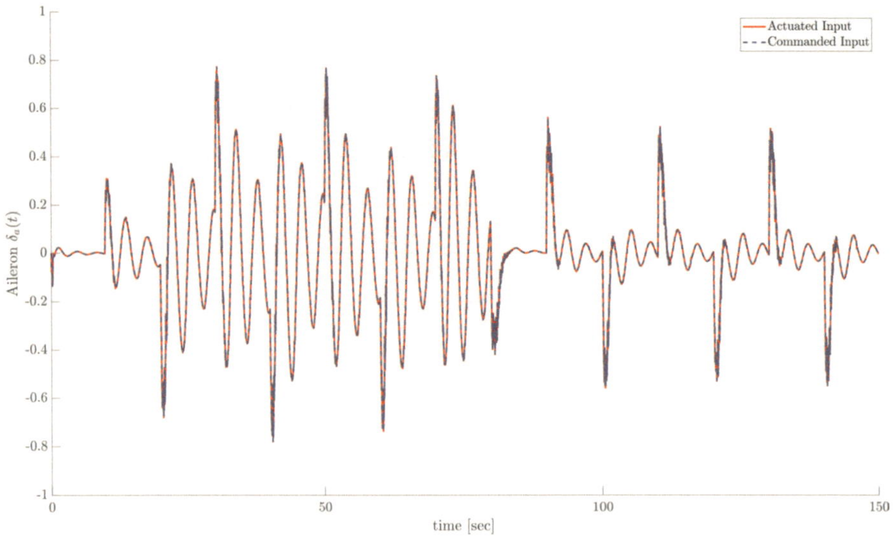

Figure 14.
Aileron control for SITL simulation with $\Gamma = \Gamma_\Theta = 2I$, $\gamma_m = 10$, $\tau_w = 30$, $\gamma_1 = 1$, and $\gamma_2 = 10$.

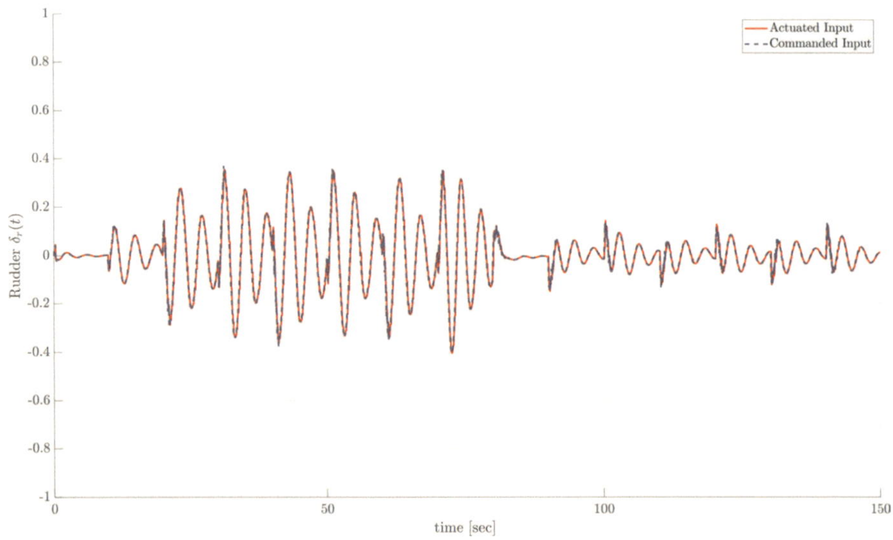

Figure 15.
Rudder control for SITL simulation with $\Gamma = \Gamma_\Theta = 2I$, $\gamma_m = 10$, $\tau_w = 30$, $\gamma_1 = 1$, and $\gamma_2 = 10$.

cancelation of the uncertainties. This is clear as the commanded control surface deflections significantly decrease with the activation of proposed adaptive controller.

Note that the uncertainty in SITL flight simulation is not guaranteed to be matched. That is, some portion of the uncertainty may not lie in the range space of input matrix B. Corresponding plant dynamics can be expressed as:

$$\dot{x} = Ax + B[u + \Delta(x)] + D\Delta_u(x) \tag{75}$$

with $B^T D = 0$ and $\Delta_u(x)$ being unmatched uncertainty. In this case, the reference model state tracking cannot be achieved since it is not possible to suppress all the effects of uncertainties with the control input. Instead, controlled output tracking can be achieved by introducing the command governor controls for the unmatched uncertainties [27].

7. Conclusions

In this chapter, a new model reference adaptive control architecture is proposed to recover the lost information in filter-based adaptive controllers such as composite/combined MRAC. In CMRAC-like formulations, the uncertainty estimation is achieved by filtering the system dynamics, which significantly improves the transient performance. However, filtering the high-frequency content in the system signals may degrade the adaptation due to lost information. This adverse effect can become even more crucial if the low-pass filter bandwidth is remarkably small. With the proposed method, the high-frequency content of the uncertainty estimation is integrated in a frequency-selective manner into the adaptive law to recover the information lost during filtering. Furthermore, learning rates in the weight update law become time-varying so that the proposed modification term behaves as a stability augmentation system. In that regard, the bandwidth of the closed-loop system is adjusted online. The efficacy of the proposed solution is illustrated through numerical simulations of aircraft roll dynamics and software-in-the-loop simulations for lateral dynamics of F-4 Phantom II fighter aircraft through the X-plane flight simulator. Furthermore, rigorous stability proof is provided by Lyapunov's stability theorem.

Author details

Metehan Yayla[1*] and Ali Turker Kutay[2]

1 Esen System Integration LLC, Ankara, Turkey

2 Department of Aerospace Engineering, Middle East Technical University, Ankara, Turkey

*Address all correspondence to: myayla@metu.edu.tr

References

[1] Boyd S, Sastry SS. Necessary and sufficient conditions for parameter convergence in adaptive control. Automatica. 1986;**22**(6):629-639

[2] Ioannou PA, Kokotovic PV. Instability analysis and improvement of robustness of adaptive control. Automatica. 1984;**20**(5):583-594

[3] Narendra KS, Annaswamy A. A new adaptive law for robust adaptation without persistent excitation. IEEE Transactions on Automatic Control. 1987;**32**(2):134-145

[4] Pomet J-B, Praly L. Adaptive nonlinear regulation: Estimation from the Lyapunov equation. IEEE Transactions on Automatic Control. 1992;**37**(6):729-740

[5] Nguyen N, Krishnakumar K, Boskovic J. An optimal control modification to model-reference adaptive control for fast adaptation. In: AIAA Guid. Navig. Control Conf. Honolulu, HI: AIAA; 2008

[6] Calise AJ, Yucelen T, Muse JA, Yang B-J. A loop recovery method for adaptive control. In: AIAA Guid. Navig. Control Conf. Chicago, IL: AIAA; 2009

[7] Yucelen T, Calise AJ. Kalman filter modification in adaptive control. Journal of Guidance, Control, and Dynamics. 2010;**33**(2):426-439

[8] Slotine J-JE, Li W. Composite adaptive control of robot manipulators. Automatica. 1989;**25**(4):509-519

[9] Volyanskyy K, Calise AJ, Yang B-J, Lavretsky E. An error minimization method in adaptive control. In: AIAA Guid. Navig. Control Conf. Keystone, CO: AIAA; 2006

[10] Adetola V, Guay M. Performance improvement in adaptive control of linearly parameterized nonlinear systems. IEEE Transactions on Automatic Control. 2010;**55**(9):2182-2186

[11] Yucelen T, De La Torre G, Johnson EN. Improving transient performance of adaptive control architectures using frequency-limited system error dynamics. International Journal of Control. 2014;**87**(11):2383-2397

[12] Höcht L, Maity A, Holzapfel F. Frequency selective learning model reference adaptive control. IET Control Theory & Applications. 2015;**9**(15): 2257-2265

[13] Na J, Yang J, Ren X, Guo Y. Robust adaptive estimation of nonlinear system with time-varying parameters. International Journal of Adaptive Control and Signal Processing. 2015; **29**(8):1055-1072

[14] Basu Roy S, Bhasin S. Novel model reference adaptive control architecture using semi-initial excitation-based switched parameter estimator. International Journal of Adaptive Control and Signal Processing. 2019; **33**(12):1759-1774

[15] Pan Y, Sun T, Yu H. On parameter convergence in least squares identification and adaptive control. International Journal of Robust and Nonlinear Control. 2019;**29**(10): 2898-2911

[16] Glushchenko AI, Petrov VA, Lastochkin KA. Adaptive control system with a variable adjustment law gain based on the recursive least squares method. Automation and Remote Control. 2021;**82**(4):619-633

[17] Yucelen T, Haddad WM. Low-frequency learning and fast adaptation in model reference adaptive control. IEEE Transactions on Automatic Control. 2012;**58**(4):1080-1085

[18] Nguyen N, Burken J, Hanson C. Optimal control modification adaptive law with covariance adaptive gain adjustment and normalization. In: AIAA Guid. Navig. Control Conf. Portland, OR: AIAA; 2011. p. 6606

[19] Yayla M, Kutay AT. A modification to adaptive control with time-varying learning rate for improved transient performance. In: AIAA SciTech Forum. San Diego, CA: AIAA; 2022

[20] Lavretsky E. Combined/composite model reference adaptive control. IEEE Transactions on Automatic Control. 2009;**54**(11):2692-2697

[21] Yang J, Na J, Gao G. Robust model reference adaptive control for transient performance enhancement. International Journal of Robust and Nonlinear Control. 2020;**30**(15): 6207-6228

[22] Lewis F, Jagannathan S, Yesildirek A. Neural Network Control of Robot Manipulators and Non-linear Systems. UK: CRC Press; 1998

[23] Park J, Sandberg IW. Universal approximation using radial basis function networks. Neural Computation. 1991;**3**:246-257

[24] Cybenko G. Approximation by superpositions of a sigmoidal function. Mathematics of Control, Signals and Systems. 1989;**2**(4):303-314

[25] Liu Q, Li D, Ge SS, Guo Y. Adaptive feedforward rbf neural network control with the deterministic persistence of excitation. Neural Computing and Applications. 2021;**33**(24):17013-17028

[26] Kim S-H, Lee H, Cho N, Kim Y. Data-efficient active weighting algorithm for composite adaptive control systems. IEEE Transactions on Automatic Control. 2022

[27] Yayla M, Arabi E, Kutay AT, Yucelen T. Command governor-based adaptive control for dynamical systems with matched and unmatched uncertainties. International Journal of Adaptive Control and Signal Processing. 2018;**32**(8):1124-1144

[28] Tao G. Adaptive Control Design and Analysis. New York, NY, USA: John Wiley & Sons, Inc.; 2003

[29] Ioannou PA, Sun J. Robust adaptive control. In: Control Theory. PTR Prentice-Hall; 1996

[30] Lavretsky E, Wise K. Robust and Adaptive Control: With Aerospace Applications. London: Advanced Textbooks in Control and Signal Processing, Springer; 2012

[31] Khalil HK. Nonlinear Systems. New York, NY: Prentice Hall; 2002

[32] Lavretsky E, Gibson TE. Projection operator in adaptive systems. arXiv preprint arXiv:1112.4232. 2011

[33] Sun J. A modified model reference adaptive control scheme for improved transient performance. IEEE Transactions on Automatic Control. 1993;**38**(8):1255-1259

[34] Gibson T, Annaswamy A, Lavretsky E. Improved transient response in adaptive control using projection algorithms and closed loop reference models. In: AIAA Guid. Navig.

Control Conf. Minneapolis, MN: AIAA; 2012. p. 4775

[35] Duarte MA, Narendra KS. Combined direct and indirect approach to adaptive control. IEEE Transactions on Automatic Control. 1989;**34**(10):1071-1075

[36] Guo K, Pan Y. Composite adaptation and learning for robot control: A survey. Annual Reviews in Control. 2023;**55**: 279-290

[37] Cho N, Shin H-S, Kim Y, Tsourdos A. Composite model reference adaptive control with parameter convergence under finite excitation. IEEE Transactions on Automatic Control. 2017;**63**(3):811-818

[38] Chowdhary G, Johnson E. A singular value maximizing data recording algorithm for concurrent learning. In: Proceedings of the American Control Conference. San Francisco, CA: IEEE; 2011. pp. 3547-3552

[39] Yayla M, Kutay AT. Guaranteed exponential convergence without persistent excitation in adaptive control. In: UKACC 11th International Conference on Control. Belfast: IEEE; 2016. pp. 1-6

[40] Miller KS. On the inverse of the sum of matrices. Mathematics Magazine. 1981;**54**(2):67-72

[41] Gorinevsky D. On the persistency of excitation in radial basis function network identification of nonlinear systems. IEEE Transactions on Neural Networks. 1995;**6**(5):1237-1244

[42] Laminar Research. X-plane: Flight simulator. X-Plane Version 11. 2022

[43] Mark D, Harold Y. Athena vortex lattice. AVL 3.6. 2022.

Chapter 4

Robust Hybrid Model Reference Adaptive Control and Output-Feedback Linearization with Applications to Quadcopter UAVs

Giri M. Kumar, Mattia Gramuglia and Andrea L'Afflitto

Abstract

This chapter presents the first robust model reference adaptive control (MRAC) system for hybrid, time-varying plants affected by parametric, matched, and unmatched uncertainties as well as uncertainties in the plant's discrete-time dynamics. This continuous-time component of this MRAC system comprises both an adaptive law and a control law that are analogous to the adaptive law and control law of classical MRAC systems. The discrete-time component of the proposed MRAC system comprises a resetting mechanism that counters the effect of resetting events in the plant dynamics. The mechanisms that guarantee robustness to unmatched uncertainties extend the well-known σ-modification and e-modification of MRAC as well as the use of continuous projection operators to a hybrid systems framework. This adaptive control framework is applied to the problem of controlling output-feedback linearized dynamical models while switching among multiple feedback-linearizing output signals according to any user-defined algorithm that is compatible with the conditions sufficient for the existence of the linearizing diffeomorphism. As an example, we solve the problem of controlling the dynamics of a quadcopter unmanned aerial vehicle (UAV) tasked with following both a user-defined trajectory and a user-defined attitude, and not just a user-defined yaw angle as it occurs in the overwhelming majority of works on this topic.

Keywords: hybrid dynamical systems, robust model reference adaptive control, output-feedback linearization, uncertain systems, quadcopters

1. Introduction

This chapter presents the first robust model reference adaptive control (MRAC) system for hybrid plants affected by parametric, matched, and unmatched uncertainties. Hybrid plants comprise dynamical models of processes that can be captured by means of both differential and difference equations. Differential equations allow describing continuous-time phenomena, whereas difference equations allow describing discrete-time phenomena. Examples of such plants include mechanical systems,

IntechOpen

whose continuous-time dynamics experience instantaneous variations due to external solicitations, elastic effects, or sudden variations in the characterizing parameters such as friction coefficients [1, 2]. Additional examples of such plants include those systems, whose dynamics are affected by continuous-time effects, both exogenous, such as disturbances, and endogenous, such as control inputs, as well as by discrete-time effects, such as decision variables drawn from a countable set of possible choices [3]. The proposed MRAC system is proven to be robust to uncertainties in both the plant's continuous-time dynamics and in its discrete-time dynamics.

The proposed results extend the results presented in [4], which propose the first MRAC system for nonlinear hybrid plants, whose dynamics are affected by matched and parametric uncertainties, to the case wherein the plant dynamics are affected by unmatched uncertainties as well. This extension has been possible by leveraging the first generalization of the LaSalle-Yoshizawa theorem to prove the pre-attractivity of compact sets for nonlinear, time-varying, hybrid systems. Furthermore, this chapter extends for the first time classical results such as the e-modification of MRAC [5], the σ-modification of MRAC [6], and the use of continuous projection operators [7] to nonlinear, time-varying, uncertain hybrid plants. Specifically, the proposed MRAC system robustifies the results in [4] with mechanisms that are analogous to the aforementioned classical robustifications of MRAC, while retaining its peculiar resetting mechanism of the reference model's dynamics. Such a mechanism, which is impossible to deduce applying classical Lyapunov-like sufficient conditions predicated assuming continuity of the system's dynamics with respect to time and Lipschitz continuity with respect to the state, allows the state of the reference model to instantaneously reduce the trajectory tracking error and ease its convergence to zero. The time at which these resetting events in the reference model occur is computed as the time at which the energy injected into the controlled system by the uncertain discrete-time dynamics exceeds the energy dissipated by the control system's continuous-time dynamics.

The application of the proposed robust hybrid MRAC system is unique and opens the way to new research ideas in the context of output-feedback linearization [8]. Indeed, the proposed adaptive control system is applied to regulate output-feedback linearized dynamical systems, whose measured output, which defines the feedback-linearizing diffeomorphism, is arbitrarily switched by the user over a countable set of alternative options. To illustrate this idea, we consider the problem of controlling a quadcopter UAV by means of an output-feedback linearizing system, which serves as a baseline controller, and a robust MRAC system to improve the tracking performance despite uncertainties and disturbances. The overwhelming literature on the control of quadcopter UAVs by means of output-feedback linearization consider only one measured output, namely, the UAV's position and yaw angle; see [9–11] for some of the latest references on this topic of a conspicuous list. To the authors' knowledge, alternative output functions, such as the UAV's position and any of the other two Euler's angles, which are commonly available for measurement using any commercial-off-the-shelf autopilot, such as those based on PX4 [12] or Ardupilot [13] to name two of the most popular ones, are not considered. The reasons for this choice substantially stem from the fact that output-feedback linearization with respect to the vehicle's position and yaw angle only requires a non-zero total thrust at all times, which is realistic in most problems of practical interest, where free fall of the UAV is not required. Output-feedback linearization with respect to the vehicle's position and either pitch or roll angle requires additional constraints on the vehicle's attitude, which do not allow hovering and pose challenges in near-equilibrium maneuvers.

Furthermore, most applications considered so far can be performed by simply tasking the UAV to follow some user-defined trajectory for its center of mass and some yaw angle. Indeed, onboard vision-based sensors, such as cameras or Lidars, are generally aligned with the UAV's roll axis and quadcopter UAVs usually operate in near-hover conditions. The proposed idea of using a hybrid MRAC system to regulate the feedback-linearized equations of motion of a quadcopter UAV allows the user to arbitrarily choose the measured output and control all six of the UAV's degrees of freedom by cycling through multiple output functions, and not only four degrees of freedom, as it occurs in existing control architectures for this class of aerial robots.

Numerical simulations prove the effectiveness of the proposed robust hybrid MRAC framework and its applicability to a variable output-feedback linearizing framework. Numerical evidence also shows how the proposed user-defined reference trajectory, yaw, pitch, and roll angles are impossible to follow without the proposed hybrid framework.

This chapter is structured as follows. In Section 2, we present the notation used in this chapter. In Section 3, we present a sufficient condition on the pre-attractivity of compact sets for nonlinear, time-varying hybrid plants. Section 4 illustrates the first key result of this chapter, namely a robust MRAC system for hybrid plants. Successively, the equations of motion of a quadcopter UAV are recalled in Section 5. Section 6 presents the second key result of this chapter, namely the application of the proposed adaptive hybrid system control framework to the feedback-linearized equations of motion of a quadcopter UAV. In Section 7, we discuss the applicability and the features of the proposed results by means of a numerical example. Finally, Section 8 draws conclusions and outlines future work directions.

2. Mathematical notation

Let $\mathbb{N}$ denote the *set of positive integers*, $\mathbb{R}$ the *set of real numbers*, $\mathbb{R}^n$ the *set of* $n \times 1$ *real column vectors*, and $\mathbb{R}^{n \times m}$ the *set of* $n \times m$ *real matrices*. The *interior of the set* $\mathcal{S} \subset \mathbb{R}^n$ is denoted by $\mathcal{S}$, the *boundary of* $\mathcal{S} \subset \mathbb{R}^n$ is denoted by $\partial\mathcal{E}$, and the *closure of* $\mathcal{S}$ is denoted by $\overline{\mathcal{S}}$. The *open ball of radius* $\rho > 0$ *centered at* $x \in \mathbb{R}^n$ is denoted by ${}_{\rho}(x)$.

The *transpose of* $B \in \mathbb{R}^{n \times m}$ is denoted by B^{T}, and the *zero vector* in $\mathbb{R}^n$ is denoted by 0_n or 0, the *zero* $n \times m$ *matrix* in $\mathbb{R}^{n \times m}$ is denoted by $0_{n \times m}$ or 0, and the *identity matrix* in $\mathbb{R}^{n \times n}$ is denoted by $\mathbf{1}_n$. The diagonal matrix whose entries are give by $x_1, \ldots, x_n$ is denoted by $\operatorname{diag}(x_1, \ldots, x_n)$. The block-diagonal matrix formed by $M_i \in \mathbb{R}^{n_i \times n_i}$, $i = 1, \ldots, p$, is denoted by $M = \operatorname{blockdiag}(M_1, \ldots, M_p)$. The *distance between the point* $x \in \mathbb{R}^n$ *and the set* $\mathcal{S}$ is denoted by $\operatorname{dist}(x, \mathcal{S})$ ([14], p. 16). We write $\|\cdot\|$ for the *Euclidean vector norm* and the corresponding *equi-induced matrix norm* ([15], Def. 9.4.1).

3. A sufficient condition on uniform pre-attractivity of compact sets

In this section, we recall elements of hybrid systems theory, which are essential to our discussion and recall the first extension of the LaSalle-Yoshizawa theorem to time-varying, nonlinear, hybrid dynamical system. Time-varying, hybrid dynamical systems can be captured by

$$\dot{x}(t) = f_c(t, x(t)), \quad (t, x(t)) \notin \mathcal{D}, \tag{1}$$

$$x(t^+) = g_d(t, x(t)), \quad (t, x(t)) \in \mathcal{D}, \tag{2}$$

with $x(t_0) = x_0$ and initial time $t_0 \in [0, \infty)$. Let $\mathcal{Z} \subseteq \mathbb{R}^n$ be an open set such that $0 \in \mathcal{Z}$. The *flow map*, $f_c : [t_0, \infty) \times \mathcal{Z} \to \mathbb{R}^n$ is Lebesgue integrable, locally bounded, and such that $f_c(t, 0_n) = 0_n$ for all $t \in [t_0, \infty)$. The *jump map* $g_d : [t_0, \infty) \times \mathcal{Z} \to \mathbb{R}^n$ is continuous and locally bounded. A *resetting event* occurs whenever $(t, x(t)) \in \mathcal{D}$ for some $t \geq t_0$. The *resetting time before* $t \geq t_0$ is defined as $t_k \triangleq \min\{t \geq t_{k-1} : (t, s(t, t_{k-1}, x_{k-1})) \notin \mathcal{D}\}$ for all $k \in \mathbb{N}$, where $s : [t_0, \infty) \times [0, \infty) \times \mathcal{Z} \to \mathcal{Z}$ denotes the *flow* of solutions of (1) and (2). The system (1) and (2) is assumed to be *left-continuous* ([16], Def. 12.1). We also assume that $(t_0, x_0) \notin \mathcal{D}$. The case whereby $(t_0, x_0) \in \mathcal{D}$ can be addressed applying similar arguments. In this chapter, we consider Krasovskii solutions of (1) and (2) [17] and make the following assumption to avoid beating, that is, to prevent solutions of (1) and (2) from incurring into the same resetting event multiple times in zero time.

Assumption 3.1 Consider the system given by (1) and (2). If $(t, x(t)) \in \overline{\mathcal{D}} \backslash \mathcal{D}$, then there exists $\varepsilon > 0$ such that, for all $\delta \in (0, \varepsilon)$, $s(t + \delta, t, x(t)) \notin \mathcal{D}$ Furthermore, if $(t_k, x(t_k)) \in \partial\mathcal{D} \cap \mathcal{D}$, then there exists $\varepsilon > 0$ such that, for all $\delta \in (0, \varepsilon)$, $s(t_k + \delta, t_k, x(t_k^+)) \notin \mathcal{D}$.

The following result provides a sufficient condition for uniform boundedness and the convergence of complete solutions of (1) and (2) to a compact set. To state this result, let $x : [t_0, \infty) \to \mathcal{Z}$ denote a solution of (1) and (2). Furthermore, let $V : [t_0, \infty) \times \mathcal{Z} \to \mathbb{R}$ be absolutely continuous over compact intervals of $[t_0, \infty)$ not containing resetting times in their interior for each $x \in \mathcal{Z}$, and, for each $t \in [t_0, \infty)$, Lipschitz continuous and regular over $\mathcal{Z}$; for the definition of regular functions and the notion of derivative of regular functions, see ([18], pp. 63–64; [19], p. 39; and [20]).

Let $W : \mathcal{Z} \to \mathbb{R}$ be absolutely continuous and nonnegative definite. Let $\bar{t}_k \in [t_0, \infty)$, $k \in \overline{\mathbb{N}}$, such that $\bar{t}_0 = t_0$, $\bar{t}_1 = t_1$, if $\sum_{j=1}^{k-1} \left[V(t_j^+, x(t_j^+)) - V(t_j, x(t_j))\right] > 0$, $k \in \mathbb{N} \backslash \{1\}$, along a solution of (1) and (2), then

$$\bar{t}_k = \inf\left\{t \in [t_0, \infty) : \int_{t_0}^{t} W(x(\tau))\mathrm{d}\tau \geq \sum_{j=1}^{k-1} \left[V(t_j^+, x(t_j^+)) - V(t_j, x(t_j))\right]\right\}, \tag{3}$$

and if $\sum_{j=1}^{k-1} \left[V(t_j^+, x(t_j^+)) - V(t_j, x(t_j))\right] \leq 0$, then $\bar{t}_k = t_1$. Finally, the *critical times* are defined as

$$\hat{t}_k \triangleq \max\{t_k, \bar{t}_k\}. \tag{4}$$

Theorem 1.1 ([4], Th. 2) Consider the hybrid, time-varying, nonlinear dynamical system given by (1) and (2), and assume that all solutions of (1) and (2) are complete. Let $V : [t_0, \infty) \times \mathcal{Z} \to \mathbb{R}$ be absolutely continuous in its first argument over compact intervals of $[t_0, \infty) \subseteq \overline{\mathbb{R}}_+$ that do not contain resetting times in their interior for each $x \in \mathcal{Z}$ and Lipschitz continuous and regular in the second argument for each $t \in [t_0, \infty)$. Assume that $\hat{t}_k \leq t_k$ for all $k \in \overline{\mathbb{N}}$, $\sum_{k=1}^{\infty} \left[V(t_k^+, x(t_k^+)) - V(t_k, x(t_k))\right]$ exists and is finite, and

$$W_1(x) \leq V(t, x) \leq W_2(x), \quad (t, x) \in [t_0, \infty) \times \mathcal{Z}, \tag{5}$$

$$\dot{V}(t,x) \leq -W(x), \ (t,x) \notin \left([t_0,\infty) \times \left(\mathcal{Z}\backslash\overline{\mathcal{A}}\right)\right) \cap \mathcal{D}, \tag{6}$$

where $W_1, W_2 : \mathcal{Z} \to \mathbb{R}$ are positive-definite, $\overline{\mathcal{A}} \subset \mathcal{Z}$ is compact and such that $0 \in \widehat{\mathcal{A}}$, and $W : \mathcal{Z} \to \mathbb{R}$ is continuously differentiable on $\mathcal{Z}\backslash\{0_n\}$, nonnegative-definite, and such that $W(x) > 0$ for all $x \in \mathcal{Z}\backslash\overline{\mathcal{A}}$. Let $r > 0$ and $c > 0$ be such that $\mathbb{B}_r(\overline{\mathcal{A}}) \subset \mathcal{Z}$ and $c < \min_{x \in \partial\mathbb{B}_r(\overline{\mathcal{A}})} W_1(x)$. If $x(t_0) \in \{x \in \mathbb{B}_r(\overline{\mathcal{A}}) : W_2(x) \leq c\}$, then every maximal solution $x(t), t \geq t_0$, of (1) and (2) is bounded uniformly in $\{t_k\}_{k \in \overline{\mathbb{N}}}$ and such that $\lim_{t\to\infty} \text{dist}\left(x(t), \overline{\mathcal{A}}\right) = 0$ uniformly in $\{t_k\}_{k \in \overline{\mathbb{N}}}$. Additionally, if $\mathcal{Z} = \mathbb{R}^n$ and both $W_1(\cdot)$ and $W_2(\cdot)$ are radially unbounded, then every maximal solution $x(\cdot)$ of (1) and (2) is uniformly bounded in $\{t_k\}_{k \in \overline{\mathbb{N}}}$ and such that $\lim_{t\to\infty} \text{dist}\left(x(t), \overline{\mathcal{A}}\right) = 0$ for all $x_0 \in \mathbb{R}^n$ uniformly in $\{t_k\}_{k \in \overline{\mathbb{N}}}$.Theorem 1.1 provides Lyapunov-like sufficient conditions on the local and global uniform pre-attractivity of the compact set $\overline{\mathcal{A}}$ ([21], Def. 7.1), that is, on the property whereby complete solutions of (1) and (2) converge to $\overline{\mathcal{A}}$. This result extends a notorious theorem for uniform ultimate boundedness ([22], Def. 4.6) of nonlinear time-varying dynamical systems that are continuous in time and Lipschitz continuous in the state vector, namely Theorem 4.18 of [22].

4. Robust model reference adaptive control for hybrid systems

This section presents the first key contribution of this chapter, namely present the first MRAC system robust to parametric, matched, and unmatched uncertainties. Section 4.1 outlines the plant and reference model dynamics. Section 4.2 presents three robust control systems, which extend the classical e-modification of MRAC [5], the σ-modification of MRAC [6], and the use of continuous projection operators [7] to hybrid plants. Finally, Section 4.3 leverages the results of Section 3 and proves the effectiveness of these control systems.

4.1 Plant and reference model dynamics

In this section, we present multiple robust MRAC schemes for nonlinear, time-varying, hybrid plants with modeling and parametric uncertainties, and uncertainties in the resetting events. To this goal, consider the *plant model*

$$\begin{aligned} \begin{bmatrix} \dot{x}(t) \\ \dot{\sigma}(t) \end{bmatrix} &= \begin{bmatrix} A_{\sigma(t)}x(t) + B_{\sigma(t)}\left[u(t) + \tilde{\Theta}^{\mathrm{T}}_{\sigma(t)}\tilde{\Phi}_{\sigma(t)}(t,x(t))\right] \\ 0 \end{bmatrix} \\ &+ \begin{bmatrix} \xi_{\sigma(t)}(t) \\ 0 \end{bmatrix}, \quad \begin{bmatrix} x(t_0) \\ \sigma(t_0) \end{bmatrix} = \begin{bmatrix} x_0 \\ \sigma_0 \end{bmatrix}, \qquad (t,x(t)) \notin \mathcal{D}_{\sigma(t)}, \end{aligned} \tag{7}$$

$$\begin{bmatrix} x(t^+) \\ \sigma(t^+) \end{bmatrix} = g_{\mathrm{d},\sigma(t)}(t,x(t)), \quad (t,x(t)) \in \mathcal{D}_{\sigma(t)}, \tag{8}$$

where $x_\sigma : [t_0,\infty) \to \mathbb{R}^n$ denotes the *plant state*, $\sigma : [t_0,\infty) \to \Sigma$ denotes the *mode*, $\Sigma \subset \mathbb{N}$ comprises the first $\sigma_{\max}$ positive integers, the piecewise continuous function $u : [t_0,\infty) \to \mathbb{R}^m$ denotes the *control input*, $A_\sigma \in \mathbb{R}^{n\times n}$, $\sigma \in \Sigma$, is unknown, the mapping $\sigma \mapsto A_\sigma$ is unknown, $B_\sigma \in \mathbb{R}^{n\times m}$ is known, the pair (A_σ, B_σ) is controllable, $\tilde{\Theta}_\sigma \in \mathbb{R}^{\tilde{N}_\sigma\times m}$ is unknown, the mapping $\sigma \mapsto \tilde{\Theta}_\sigma$ is unknown, the *regressor vector* $\tilde{\Phi}_\sigma : [t_0, \infty) \times \mathbb{R}^n \to \mathbb{R}^{\tilde{N}_\sigma}$ is Lipschitz continuous and known, $\xi_\sigma : [t_0, \infty) \to \mathbb{R}^n$ is unknown, piecewise continuous, and such that $\|\xi_\sigma(t)\| \leq \xi_{\sigma,\max}$, $\xi_{\sigma,\max} \geq 0$ is known, the mapping $\sigma \mapsto \xi_\sigma(\cdot)$ is unknown, and the mapping $\sigma \mapsto \xi_{\sigma,\max}$ is known. The i-th resetting time of the resetting event $\mathcal{D}_{\sigma_{i-1}}$, $(i, \sigma) \in \mathbb{N} \times \Sigma$ is given by

$$t_{\text{plant},i} = \min\{t > t_{\text{plant},i-1} : (t, s_{\sigma_{i-1}}(t, t_{\text{plant},i-1}, x_{i-1})) \notin \mathcal{D}_{\sigma_{i-1}}\}, \tag{9}$$

where $\{\sigma_{i-1}\}_{i\in\mathbb{N}} \subset \Sigma$, $s_{\sigma_{i-1}}(\cdot, t_{\text{plant},i-1}, x_{i-1})$ denotes the flow of the plant (7) and (8) originated by the $\mathcal{D}_{\sigma_{i-1}}$ resetting event at time $t_{\text{plant},i-1}$ and from the initial condition x_{i-1}. In this chapter, the jump map $g_{\mathrm{d},\sigma}(\cdot,\cdot)$, $\sigma \in \Sigma$, is known and uncontrollable, and the resetting events $\{\mathcal{D}_\sigma\}_{\sigma\in\Sigma} \subset [t_0, \infty) \times \mathbb{R}^n$ are unknown. We also assume that, with any piecewise continuous control input $u(\cdot)$, (7) and (8) verify Assumption 3.1. In problems involving mechanical systems subject to elastic collisions, Assumption 3.1 is verified if collisions do not occur in arbitrarily small time intervals, which is a realistic modeling assumption.

Next, consider the *reference model*

$$\begin{bmatrix} \dot{x}_{\text{ref}}(t) \\ \dot{\sigma}(t) \end{bmatrix} = \begin{bmatrix} A_{\text{ref},\sigma(t)} x_{\text{ref}}(t) + B_{\text{ref},\sigma(t)} r(t) \\ 0 \end{bmatrix}, \quad \begin{bmatrix} x_{\text{ref}}(t_0) \\ \sigma(t_0) \end{bmatrix} = \begin{bmatrix} x_{\text{ref},0} \\ \sigma_0 \end{bmatrix}, \quad (t, x_{\text{ref}}(t)) \notin \mathcal{D}_{\text{ref},\sigma(t)}, \tag{10}$$

$$\begin{bmatrix} x_{\text{ref}}(t^+) \\ \sigma(t^+) \end{bmatrix} = g_{\mathrm{d,ref},\sigma(t)}(t, x_{\text{ref}}(t)), \quad (t, x_{\text{ref}}(t)) \in \mathcal{D}_{\text{ref},\sigma(t)}, \tag{11}$$

where $A_{\text{ref},\sigma} \in \mathbb{R}^{n\times n}$, $\sigma \in \Sigma$, is Hurwitz and such that

$$A_{\text{ref},\sigma} = A_\sigma + B_\sigma K_{x,\sigma}^{\mathrm{T}} \tag{12}$$

for some $K_{x,\sigma} \in \mathbb{R}^{n\times m}$, $B_{\text{ref},\sigma} \in \mathbb{R}^{n\times m}$ is such that

$$B_{\text{ref},\sigma} = B_\sigma K_{r,\sigma}^{\mathrm{T}} \tag{13}$$

for some $K_{r,\sigma} \in \mathbb{R}^{m\times m}$, and the *reference command input* $r : [t_0, \infty) \to \mathbb{R}^m$ is piecewise continuous and bounded. The reference model's dynamics capture the desired closed-loop system's dynamics. The jump map $g_{\mathrm{d,ref},\sigma}(\cdot,\cdot)$ and the set of resetting events $\{\mathcal{D}_{\text{ref},\sigma}\}_{\sigma\in\Sigma}$ are presented in the following. The *matching conditions* (12) and (13) signify that, for each mode, the reference model dynamics can be mimicked by the controlled plant dynamics.

4.2 Control system outline

Our goal is to derive adaptive control laws to steer the trajectories of (7) and (8) toward the trajectories of the reference model (10) and (11), despite uncertainties in the plant model. To this goal, let

$$e(t) \triangleq x(t) - x_{\text{ref}}(t), \quad t \geq t_0, \tag{14}$$

denote the *trajectory tracking error*. Furthermore, define

$$\Phi_\sigma(t,x) \triangleq \left[x^{\mathrm{T}}, r^{\mathrm{T}}(t), -\tilde{\Phi}_\sigma^{\mathrm{T}}(t,x),\right]^{\mathrm{T}}, \qquad (\sigma,t,x) \in \Sigma \times [t_0,\infty) \times \mathbb{R}^n, \tag{15}$$

$$\Theta_\sigma \triangleq \left[K_{x,\sigma}^{\mathrm{T}}, K_{r,\sigma}^{\mathrm{T}}, \tilde{\Theta}_\sigma^{\mathrm{T}}\right]^{\mathrm{T}}, \tag{16}$$

$$\overline{\Phi}_\sigma(t,x) \triangleq \left[\chi_{\Sigma_1}(\sigma)\Phi_1^{\mathrm{T}}(t,x), \ldots, \chi_{\Sigma_p}(\sigma)\Phi_p^{\mathrm{T}}(t,x)\right]^{\mathrm{T}}, \tag{17}$$

$$\Theta \triangleq \left[\Theta_1^{\mathrm{T}}, \ldots, \Theta_p^{\mathrm{T}}\right]^{\mathrm{T}}, \tag{18}$$

and $N \triangleq 2n + m + \sum_{\sigma=1}^{p} N_\sigma$, and note that

$$\begin{bmatrix} \dot{e}(t) \\ \dot{\sigma}(t) \end{bmatrix} = \begin{bmatrix} A_{\mathrm{ref},\sigma(t)} e(t) + B_{\sigma(t)}\left[u(t) - \Theta^{\mathrm{T}}\overline{\Phi}_{\sigma(t)}(t,x(t))\right] \\ 0 \end{bmatrix} + \begin{bmatrix} \xi_{\sigma(t)}(t) \\ 0 \end{bmatrix}, \quad \begin{bmatrix} e(t_0) \\ \sigma(t_0) \end{bmatrix} = \begin{bmatrix} x_0 - x_{\mathrm{ref},0} \\ \sigma_0 \end{bmatrix},$$
$$\left((t,x(t)) \notin \mathcal{D}_{\sigma(t)}\right) \wedge \left((t,x_{\mathrm{ref}}(t)) \notin \mathcal{D}_{\mathrm{ref},\sigma(t)}\right), \tag{19}$$

$$\begin{bmatrix} e(t^+) \\ \sigma(t^+) \end{bmatrix} = g_{\mathrm{d},\sigma(t)}(t,x(t)) - g_{\mathrm{d,ref},\sigma(t)}(t,x_{\mathrm{ref}}(t)),$$
$$\left((t,x(t)) \in \mathcal{D}_{\sigma(t)}\right) \vee \left((t,x_{\mathrm{ref}}(t)) \in \mathcal{D}_{\mathrm{ref},\sigma(t)}\right), \tag{20}$$

where $\wedge$ represents the operator *and*, $\vee$ represents the operator *or*, $\Sigma_j \subseteq \Sigma$ denotes the j-th of $\sigma_{\max}$ partitions of Σ, and $\chi_{\Sigma_j} : \Sigma_j \to \{0,1\}$ symbolizes the *indicator function*.

Remark 4.1 The hybrid dynamical systems given by (7) and (8), (10) and (11), and (19) and (20) can be reduced to the same form as (1) and (2) by proceeding as in [17].To pursue our goal, consider also the *control law*

$$\eta\left(\hat{\Theta}, \overline{\Phi}_\sigma(t,x)\right) = \hat{\Theta}^{\mathrm{T}}\overline{\Phi}_\sigma(t,x), \quad \left(\sigma,t,x,\hat{\Theta}\right) \in \Sigma \times [t_0,\infty) \times \mathbb{R}^n \times \mathbb{R}^{N\times m}, \tag{21}$$

and the *adaptive laws*

$$\dot{\hat{\Theta}}(t) = \hat{\Theta}_{\mathrm{d}}(t) - \gamma_{\sigma(t)}\hat{\Theta}(t) \ (\sigma\text{-mod.}), \tag{22}$$

$$\dot{\hat{\Theta}}(t) = \hat{\Theta}_{\mathrm{d}}(t) - \gamma_{\sigma(t)}\left\|e^{\mathrm{T}}P_{\sigma(t)}B_{\sigma(t)}\right\|\hat{\Theta}(t) \ (e\text{-mod.}), \tag{23}$$

$$\dot{\hat{\Theta}}(t) = \mathrm{Proj}_{\sigma(t)}\left(\hat{\Theta}(t), \hat{\Theta}_{\mathrm{d}}(t)\right) \quad (\text{proj. operator}), \tag{24}$$

with

$$\hat{\Theta}_{\mathrm{d}}(t) \triangleq -\Gamma_{\sigma(t)}\overline{\Phi}_{\sigma(t)}(t,x(t))e^{\mathrm{T}}(t)P_{\sigma(t)}B_{\sigma(t)}, \tag{25}$$

$\hat{\Theta}(t_0) = \hat{\Theta}_0$, and $t \geq t_0$. These adaptive laws are to be considered as *alternatives* to one another, and capture extensions to hybrid systems of the the σ-modification of MRAC [6], the e-modification of MRAC [5], and the projection operator [23], respectively. In (22)–(24), the *adaptive rate matrix* $\Gamma_\sigma \in \mathbb{R}^{N\times N}$, $\sigma \in \Sigma$, is positive-definite, $\gamma_\sigma > 0$, $P_\sigma \in \mathbb{R}^{n\times n}$ is the positive-definite and such that

$$0_{n\times n} = A_{\mathrm{ref},\sigma}^{\mathrm{T}} P_\sigma + P_\sigma A_{\mathrm{ref},\sigma} + Q_\sigma, \tag{26}$$

and $Q_\sigma \in \mathbb{R}^{n\times n}$ is user-defined and positive-definite. To define the *matrix projection operator* in (24), firstly consider the definition of vector projection operator.

Definition 4.1 Let $\mathcal{X} \subseteq \mathbb{R}^n$ be convex, and let $h_\sigma : \mathcal{X} \to \mathbb{R}$, $\sigma \in \Sigma$, denote a continuously differentiable convex function over $\mathcal{X}$ such that $\inf_{x\in\mathcal{X}} h_\sigma(x) < 0$. The *vector projection operator induced* by $h_\sigma(\cdot)$ over $\mathcal{X}$ is defined as $\mathrm{proj}_\sigma : \mathcal{X} \times \mathbb{R}^n \to \mathbb{R}^n$, $\sigma \in \Sigma$, such that if $(x, x_\mathrm{d}) \in \mathcal{S}_\sigma$, then

$$\mathrm{proj}_\sigma(x, x_\mathrm{d}) \triangleq x_\mathrm{d} - h_\sigma(x) \frac{\left(\frac{\partial h_\sigma(x)}{\partial x}\right)^{\mathrm{T}} \frac{\partial h_\sigma(x)}{\partial x}}{\frac{\partial h_\sigma(x)}{\partial x} \left(\frac{\partial h_\sigma(x)}{\partial x}\right)^{\mathrm{T}}} x_\mathrm{d}, \tag{27}$$

and if $(x, x_\mathrm{d}) \notin \mathcal{S}_\sigma$, then

$$\mathrm{proj}_\sigma(x, x_\mathrm{d}) \triangleq x_\mathrm{d}, \tag{28}$$

where $\mathcal{S}_\sigma \triangleq \left\{ (x, x_\mathrm{d}) \in \mathcal{X} \times \mathbb{R}^n : h_\sigma(x) > 0, \frac{\partial h_\sigma(x)}{\partial x} x_\mathrm{d} > 0 \right\}$ and $\sigma \in \Sigma$.

Definition 4.2 ([24], Ch. 11) Let $\mathcal{X}_i \subseteq \mathbb{R}^N$ be convex, $i = 1, \dots, m$, and let $h_{\sigma,i} : \mathcal{X}_i \to \mathbb{R}$, $\sigma \in \Sigma$, denote a continuously differentiable convex function over $\mathcal{X}_i$ such that $\inf_{\pi\in\mathcal{X}_i} h_{\sigma,i}(\pi) < 0$. The *matrix projection operator induced by* $h_{\sigma,i}(\cdot)$, $(\sigma, i) \in \Sigma \times \{1, \dots, m\}$, *over* $\prod_{j=1}^m \mathcal{X}_i$ is defined as $\mathrm{Proj}_\sigma : \left(\prod_{i=1}^m \mathcal{X}_i\right) \times \mathbb{R}^{N\times m} \to \mathbb{R}^{N\times m}$ such that

$$\begin{aligned} \mathrm{Proj}_\sigma(X, X_\mathrm{d}) &= \left[\mathrm{proj}_\sigma(x_1, x_{\mathrm{d},1}), \dots, \mathrm{proj}_\sigma(x_m, x_{\mathrm{d},m})\right], \\ (X, X_\mathrm{d}) &\in \left(\prod_{i=1}^m \mathcal{X}_i\right) \times \mathbb{R}^{N\times m}, \end{aligned} \tag{29}$$

where $X = [x_1, \dots, x_m]$ and $X_\mathrm{d} = [x_{\mathrm{d},1}, \dots, x_{\mathrm{d},m}]$.The functions $h_{\sigma,i}(\cdot)$, $(\sigma, i) \in \Sigma \times \{1, \dots, m\}$, employed to define the vector projection operator, and, hence, the matrix projection operator must be chosen carefully. Indeed, for each $\sigma \in \Sigma$ and for all $i \in \{1, \dots, m\}$, the solution of

$$\dot{X}(t) = \mathrm{Proj}_\sigma\left(X(t), \dot{X}(t)\right), \qquad X(t_0) = X_0, \qquad t \geq t_0, \tag{30}$$

is such that $x_i(t) \in \overline{\Omega}_{\sigma,i,1}$ for all $t \geq t_0$, where

$$\overline{\Omega}_{\sigma,i,1} \triangleq \{x_i \in \mathcal{X}_i : h_{\sigma,i}(x) \leq 1\}. \tag{31}$$

Thus, $h_{\sigma,i}(\cdot)$ must be chosen so that, for each $k \in \mathbb{N}$ and for each $i \in \{1, \dots, m\}$, $\hat{\theta}_i(t_k) \in \overline{\Omega}_{\sigma(t_k^+),1}$, where $\hat{\theta}_i(\cdot)$ denotes the i-th column of $\hat{\Theta}(\cdot)$.

Next, consider the *Lyapunov function candidate*

$$V(t, e, \Delta\Theta) \triangleq e^{\mathrm{T}} P_{\sigma(t)} e + \mathrm{tr}\left(\Delta\Theta^{\mathrm{T}} \Gamma^{-1} \Delta\Theta\right), \quad (t, \varepsilon, \Delta\Theta) \in [t_0, \infty) \times \mathbb{R}^n \times \mathbb{R}^{N\times m}, \tag{32}$$

where $\Delta\Theta(t)\triangleq\hat{\Theta}(t) - \Theta$, and we define

$$W(e)\triangleq\overline{\lambda}_{\min}\left(\{Q_\sigma\}_{\sigma\in\Sigma}\right)\|e\|^2, \quad e\in\mathbb{R}^n, \tag{33}$$

where $\overline{\lambda}_{\min}\left(\{Q_\sigma\}_{\sigma\in\Sigma}\right)\triangleq\min\{\lambda_{\min}(Q_\sigma), \sigma\in\Sigma\}$. Each of the adaptive laws (22)–(24) are switched dynamical systems, that is, they experience discontinuities in their dynamics, but not in their state matrices. Thus, the adaptive gains are computed as Carathéodory continuous solutions of (22)–(24). Thus, discontinuities of $V\left(t, e(t), \hat{\Theta}(t)\right), t\geq t_0$, are exclusively due to discontinuities in $e^{\mathrm{T}}(t)P_{\sigma(t)}e(t)$.

The set of resetting events of the reference model are defined as $\mathcal{D}_{\mathrm{ref},\sigma_{i_w}}\triangleq\{t_{\mathrm{ref},i_w}\}\times\mathbb{R}^n$, $(i,w)\in\mathbb{N}\times\mathbb{N}$, where

$$\begin{aligned} t_{\mathrm{ref},i_w}\triangleq\inf\Big\{t>\max\{t_{\mathrm{plant},i}, t_{\mathrm{ref},i_w-1}\} : \int_{t_0}^{t} W(e(\tau))\mathrm{d}\tau \\ \geq\sum_{j=1}^{k-1}\left[V\left(t_j^+, e\left(t_j^+\right), \hat{\Theta}(t_j)\right) - V\left(t_j, e(t_j), \hat{\Theta}(t_j)\right)\right]\Big\}, \end{aligned} \tag{34}$$

and k designates the generic index for resetting times. Thus, we partition the set of resetting times of (19) and (20) as $\{t_k\}_{k\in\mathbb{N}} = \{t_{\mathrm{plant},i}\}_{i\in\mathbb{N}}\cup_{i\in\mathbb{N}}\{t_{\mathrm{tran},i_w}\}_{w\in\mathbb{N}}$. The jump maps $g_{\mathrm{d,ref},\sigma}(t, x_{\mathrm{ref}})$ $(\sigma, t, x_{\mathrm{ref}})\in\Sigma\times[t_0,\infty)\times\mathbb{R}^n$, are such that

$$\begin{aligned} x_{\mathrm{ref}}\left(t_{\mathrm{ref},i_w}^+\right) = x\left(t_{\mathrm{ref},i_w}\right) - \sqrt{\frac{e^{\mathrm{T}}\left(t_{\mathrm{ref},i_w}\right)P_{\sigma\left(t_{\mathrm{ref},i_w}\right)}e\left(t_{\mathrm{ref},i_w}\right) - z_{\mathrm{ref},i_w}}{h_{\mathrm{ref}}^{\mathrm{T}}\left(t_{\mathrm{ref},i_w}, e\left(t_{\mathrm{ref},i_w}\right)\right)h_{\mathrm{ref}}\left(t_{\mathrm{ref},i_w}, e\left(t_{\mathrm{ref},i_w}\right)\right)}} \\ \cdot P^{-\frac{1}{2}}_{\sigma\left(t_{\mathrm{ref},i_w}^+\right)}h_{\mathrm{ref}}\left(t_{\mathrm{ref},i_w}, e\left(t_{\mathrm{ref},i_w}\right)\right), \qquad (i,w)\in\mathbb{N}\times\mathbb{N}, \end{aligned} \tag{35}$$

where $h_{\mathrm{ref}} : [t_0,\infty)\times\mathbb{R}^n\to\mathbb{R}^M$ is such that

$$h_{\mathrm{ref}}^{\mathrm{T}}(t,\varepsilon)h_{\mathrm{ref}}(t,\varepsilon)>0, \quad (t,\varepsilon)\in[t_0,\infty)\times\mathbb{R}^n\backslash\{0\}, \tag{36}$$

$z_{\mathrm{ref},i_w}\in\left(0, e^{\mathrm{T}}\left(t_{\mathrm{ref},i_w}\right)P_{\sigma\left(t_{\mathrm{ref},i_w}\right)}e\left(t_{\mathrm{ref},i_w}\right)\right)$ and is user-defined, the series $\sum_{i=1}^{\infty}\sum_{w=1}^{\infty}z_{\mathrm{ref},i_w}$ is convergent, and $P_\sigma^{\frac{1}{2}}\in\mathbb{R}^{n\times n}$, $\sigma\in\Sigma$, is symmetric, positive-definite, and such that $P_\sigma = P_\sigma^{\frac{1}{2}}P_\sigma^{\frac{1}{2}}$. An interpretation of the resetting time (34) is the following. This is the time at which the energy injected into the controlled system by the uncertain discrete-time dynamics exceeds the energy dissipated by the control system's continuous-time dynamics.

To improve the closed-loop trajectory tracking error dynamics at isolated time instants, consider the user-defined time instants $\cup_{i,w\in\mathbb{N}}\{\tilde{t}_{\mathrm{ref},i_w}\}$, where $\tilde{t}_{\mathrm{ref},i_w}>\max\{t_{\mathrm{plant},i}, t_{\mathrm{ref},i_w-1}\}$, $(i,w)\in\mathbb{N}\times\mathbb{N}$, and set

$$x_{\mathrm{ref}}\left(\tilde{t}_{\mathrm{ref},i_w}^+\right) = x\left(\tilde{t}_{\mathrm{ref},i_w}\right), \quad (i,w)\in\mathbb{N}\times\mathbb{N}. \tag{37}$$

Rearranging the indexes of the plant's resetting events, these user-defined resetting times will be considered resetting times of the plant, that is, we will set $\cup_{i\in\mathbb{N}}\cup_{w\in\mathbb{N}}\{\tilde{t}_{\mathrm{tran},i_w}\}\subset\cup_{i\in\mathbb{N}}\{t_{\mathrm{plant},i}\}$.

4.3 Main result

The effectiveness of the control law (21) and of the three alternative adaptive laws (22)–(24) is captured by the following result. For the statement of this result, if we employ the adaptive laws (22) or (23), then let

$$\overline{\mathcal{A}} = \{(e, \Delta\Theta) : \|e\| \leq c_e, \|\Delta\Theta\|_{\mathrm{F}} \leq c_{\Delta\Theta}\}, \tag{38}$$

where $\Delta\Theta(t) \triangleq \hat{\Theta}(t) - \Theta, t \geq t_0$. Alternatively, if we employ the adaptive law (24), then let

$$\overline{\mathcal{A}} = \left\{(e, \hat{\Theta}) : \|e\| \leq c_e, \max_{\sigma \in \Sigma} h_\sigma\left(\mathrm{col}_j\left(\hat{\Theta}\right)\right) \leq 1, j = 1, \ldots, m\right\}. \tag{39}$$

Expressions for c_e and $c_{\Delta\Theta}$ are omitted for brevity, and can be deduced by proceeding as in ([24], p. 325).

Theorem 1.2 Consider the trajectory tracking error dynamics (19) and (20), the control law (21), the adaptive laws (22)–(24), and the reference model (10) and (11). Assume that $u(t) = \eta\left(\hat{\Theta}(t), \overline{\Phi}_\sigma(t, x(t))\right), t \geq t_0$ and the matching conditions (12) and (13) are verified. Additionally, if using the adaptive law (24), assume that $\theta_i \in \overline{\Omega}_{\sigma,i,1}$ for all $(\sigma, i) \in \Sigma \times \{1, \ldots, m\}$, where θ_i denotes the i-th column of Θ given by (18) and $\overline{\Omega}_{\sigma,i,1}$ is defined in (31). Then, both the trajectory tracking error $e(\cdot)$ and the adaptive gain matrix $\hat{\Theta}(\cdot)$ are bounded uniformly in $\{t_k\}_{k \in \overline{\mathbb{N}}}$. Furthermore, there exists a compact set $\overline{\mathcal{A}}$ given by (38) when using (22) or (23) or given by (39) when using (24) such that $\lim_{t \to \infty} \mathrm{dist}\left((e(t), \Delta\Theta(t)), \overline{\mathcal{A}}\right) = 0$.

Proof: Only the key passages of this proof are presented for brevity. The Lyapunov function candidate (32) is such that

$$W_1\left(e, \hat{\Theta}\right) \leq V\left(t, e, \hat{\Theta}\right) \leq W_2\left(e, \hat{\Theta}\right), \qquad (t, e, \Delta\Theta) \in [t_0, \infty) \times \mathbb{R}^n \times \mathbb{R}^{N \times N}, \tag{40}$$

where

$$W_1\left(e, \hat{\Theta}\right) \triangleq \overline{\lambda}_{\min}\left(\{P_\sigma\}_{\sigma \in \Sigma}\right) \|e\|^2 + \mathrm{tr}\left(\Delta\Theta^{\mathrm{T}}\Gamma^{-1}\Delta\Theta\right),$$
$$W_2\left(e, \hat{\Theta}\right) \triangleq \overline{\lambda}_{\max}\left(\{P_\sigma\}_{\sigma \in \Sigma}\right) \|e\|^2 + \mathrm{tr}\left(\Delta\Theta^{\mathrm{T}}\Gamma^{-1}\Delta\Theta\right)$$

are radially unbounded. Thus, following classical arguments such as those exposed in ([24], Ch. 11) or ([25], Ch. 8) for each of the adaptive laws (22)-(24), we can prove that $\dot{V}\left(t, e(t), \hat{\Theta}(t)\right) < 0$ for all $\left(e, \hat{\Theta}\right) \notin \overline{\mathcal{A}}$, where $\overline{\mathcal{A}}$ is compact and such that $0 \in \mathcal{A}$.

Next, proceeding as in the proof of Theorem 4 in [4], we can prove that Assumption 3.1 is verified by (19) and (20) and

$$\sum_{k=1}^{\infty}\left[V\left(t_k^+, e\left(t_k^+\right), \hat{\Theta}(t_k)\right) - V\left(t_k, e(t_k), \hat{\Theta}(t_k)\right)\right]$$

exists and is finite. Thus, Theorem 1.1 implies that maximal solutions of (19) and (20) and of (22)-(24) are uniformly bounded in $\{t_k\}_{k \in \overline{\mathbb{N}}}$, and $\lim_{t \to \infty} \mathrm{dist}\left(x(t), \overline{\mathcal{A}}\right) = 0$ for all $\left(e_0, \hat{\Theta}_0\right) \in \mathbb{R}^n \times \mathbb{R}^{N \times m}$ uniformly in $\{t_k\}_{k \in \overline{\mathbb{N}}}$. ■

5. Equations of motion of a quadcopter UAV

In this section, we present the equations of motion of a quadcopter UAV. To this goal, let UAV's *mass* be denoted by $m > 0$, let the UAV's *matrix of inertia* be captured by the diagonal, positive-definite matrix $I \triangleq \mathrm{diag}(I_{11}, I_{22}, I_{33}) \in \mathbb{R}^{3\times3}$, and let the gravitational acceleration be denoted by $g > 0$. Finally, let the *UAV's position* be captured by $r : [t_0, \infty) \to \mathbb{R}^3$, the *UAV's roll angle* be denoted by $\phi : [t_0, \infty) \to \left(-\frac{\pi}{2}, \frac{\pi}{2}\right)$, the *UAV's pitch angle* be denoted by $\theta : [t_0, \infty) \to \left(-\frac{\pi}{2}, \frac{\pi}{2}\right)$, the *UAV's yaw angle* be denoted by $\psi : [t_0, \infty) \to [0, 2\pi)$, the *UAV's velocity* with respect to the inertial reference frame $\mathbb{I}$ be denoted by $v : [t_0, \infty) \to \mathbb{R}^3$, the UAV's *angular velocity* with respect to $\mathbb{I}$ be denoted by $\omega : [t_0, \infty) \to \mathbb{R}^3$, and the the UAV's *state vector* be denoted by $x(t) \triangleq \left[r^{\mathrm{T}}(t), \phi(t), \theta(t), \psi(t), v^{\mathrm{T}}(t), \omega^{\mathrm{T}}(t)\right]^{\mathrm{T}}$. Note that the UAV's state vector is usually readily available by employing any commercial-off-the-shelf autopilot system such as those based on PX4 [12] or Ardupilot [13].

Neglecting the inertial counter-torque and the gyroscopic effect [26], the UAV's continuous-time dynamics are given by

$$\dot{r}(t) = v(t), \quad r(t_0) = r_0, \quad t \in [t_0, \infty), \tag{41}$$

$$\begin{aligned}\dot{v}(t) = {} & \frac{1}{m} R(\phi(t), \theta(t), \psi(t))[0, 0, u_1(t)]^{\mathrm{T}} - [0, 0, g]^{\mathrm{T}} \\ & - \frac{1}{2m} \rho S R^{\mathrm{T}}(\phi(t), \theta(t), \psi(t)) C_{\mathrm{D}} \|v(t)\| v(t), \qquad v(t_0) = v_0,\end{aligned} \tag{42}$$

$$\begin{bmatrix} \dot{\phi}(t) \\ \dot{\theta}(t) \\ \dot{\psi}(t) \end{bmatrix} = \Gamma^{-1}(\phi(t), \theta(t))\omega(t), \qquad \begin{bmatrix} \phi(t_0) \\ \theta(t_0) \\ \psi(t_0) \end{bmatrix} = \begin{bmatrix} \phi_0 \\ \theta_0 \\ \psi_0 \end{bmatrix}, \tag{43}$$

$$\dot{\omega}(t) = I^{-1}\left(\begin{bmatrix} u_2(t) \\ u_3(t) \\ u_4(t) \end{bmatrix} - \omega^{\times}(t) I \omega(t) \right), \qquad \omega(t_0) = \omega_0, \tag{44}$$

where the rotation matrix

$$\begin{aligned} R(\phi, \theta, \psi) \triangleq {} & \begin{bmatrix} \cos\psi & -\sin\psi & 0 \\ \sin\psi & \cos\psi & 0 \\ 0 & 0 & 1 \end{bmatrix} \begin{bmatrix} \cos\theta & 0 & \sin\theta \\ 0 & 1 & 0 \\ -\sin\theta & 0 & \cos\theta \end{bmatrix} \begin{bmatrix} 1 & 0 & 0 \\ 0 & \cos\phi & -\sin\phi \\ 0 & \sin\phi & \cos\phi \end{bmatrix}, \\ & (\phi, \theta, \psi) \in \left(-\frac{\pi}{2}, \frac{\pi}{2}\right) \times \left(-\frac{\pi}{2}, \frac{\pi}{2}\right) \times [0, 2\pi), \end{aligned} \tag{45}$$

captures the UAV's attitude relative to the inertial reference frame $\mathbb{I}$ ([27], Ch. 1), $\rho > 0$ captures the air density, which is considered unknown, $S > 0$ captures the UAV's cross section area, which is considered unknown, $C_{\mathrm{D}} \in \mathbb{R}^{3\times3}$ is diagonal, positive-definite, captures the UAV's drag coefficients, and is unknown, and

$$\Gamma(\phi, \theta) \triangleq \begin{bmatrix} 1 & 0 & -\sin\theta \\ 0 & \cos\phi & \cos\theta\sin\phi \\ 0 & -\sin\phi & \cos\theta\cos\phi \end{bmatrix}. \tag{46}$$

We recall that $\Gamma(\phi, \theta)$ is invertible for all $(\phi, \theta) \in \left(-\frac{\pi}{2}, \frac{\pi}{2}\right) \times \left(-\frac{\pi}{2}, \frac{\pi}{2}\right)$ ([27], Ch. 1). The *total thrust force* produced by the UAV's propellers is defined as

$$u_1(t) \triangleq [1, 0]\delta(t), \qquad t \in [t_0, \infty), \tag{47}$$

where

$$\dot{\delta}(t) = \begin{bmatrix} 0 & \tau^{-1} \\ 0 & 0 \end{bmatrix} \delta(t) + \begin{bmatrix} 0 \\ J^{-1} \end{bmatrix} \eta_1(t), \quad \begin{bmatrix} u_1(t_0) \\ \dot{u}_1(t_0) \end{bmatrix} = \begin{bmatrix} u_{1,0} \\ u_{1,0,\mathrm{d}} \end{bmatrix}, \tag{48}$$

captures the motors' dynamics, $\tau > 0$ denotes a time constant, $J > 0$ captures the motors' inertia, and $\eta_1 : [0, \infty) \to \mathbb{R}$ denotes the *total thrust force's virtual control input*. The *roll moment* produced by the UAV's propellers is denoted by $u_2(\cdot)$, the *pitch moment* produced by the UAV's propellers is denoted by $u_3(\cdot)$, and the *yaw moment* produced by the UAV's propellers is denoted by $u_4(\cdot)$. The *UAV's control input* is defined as $u(t) \triangleq [u_1(t), u_2(t), u_3(t), u_4(t)]^{\mathrm{T}}$, $t \in [t_0, \infty)$, and the *vector of thrust forces* produced by each propeller is defined as

$$T(t) \triangleq M_{T,u} u(t), \qquad t \in [t_0, \infty), \qquad k \in \{0, \ldots, n_{\mathrm{w}} - 1\}, \tag{49}$$

where the ith component of $T(\cdot)$, $i = 1, \ldots, 4$, namely $T_i : [0, \infty) \to [t_0, \infty)$, denotes the thrust force produced by the ith propeller, $M_{T,u} \triangleq \frac{1}{4} \begin{bmatrix} 1 & 0 & 2l^{-1} & -c_{\mathrm{T}}^{-1} \\ 1 & -2l^{-1} & 0 & c_{\mathrm{T}}^{-1} \\ 1 & 0 & -2l^{-1} & -c_{\mathrm{T}}^{-1} \\ 1 & 2l^{-1} & 0 & c_{\mathrm{T}}^{-1} \end{bmatrix}$, $l > 0$ denotes the distance of the propellers from the vehicle's barycenter, and $c_{\mathrm{T}} > 0$ denotes the propellers' drag coefficient [26].

Quadcopter UAVs are under-actuated and, in particular, only four of their six degrees of freedom can be controlled directly [26]. In this chapter, we are interested in steering the UAV's position and attitude along user-defined reference trajectories by controlling the UAV's position and cyclically controlling at high frequency one of the three Euler angles $\phi(\cdot)$, $\theta(\cdot)$, and $\psi(\cdot)$ at the time.

6. Output-feedback linearization of multi-rotor UAVs

In this section, we discuss the output-feedback linearization problem of the plant model given by (41)–(44) and (48). Specifically, in Sections 6.1, 6.2, and 6.3, we discuss the output-feedback linearization problem employing the UAV position and yaw angle, the UAV position and pitch angle, and the UAV position and roll angle as measured outputs, respectively. In Section 6.4, we unify the framework presented in Sections 6.1–6.3 and illustrate how the problem of controlling the output-feedback linearized dynamics can be reduced to the problem of controlling an MRAC system. In Section 6.5, which presents the key result of this chapter, we apply the MRAC framework for hybrid plants presented in Section 4 to control a multi-rotor UAV, such as a quadcopter or an X8-copter. As already remarked in Section 1, this result is groundbreaking because, thus far, the control of multi-rotor UAVs by means of output-feedback linearization allows to impose the reference trajectory for the vehicle's position and only one of the three angles that capture its attitude.

6.1 Feedback linearization relative to position and yaw angle

To feedback-linearize (41)–(44) relative to the vehicle's position vector and yaw angle, we set $z_3(t) \triangleq [r^{\mathrm{T}}(t), \psi(t)]^{\mathrm{T}}$, $t \in [t_0, \infty)$, as a *linearizing output*, and applying Proposition 5.1.2 of [8], and we verify that the dynamical system given by (41)–(44) and (48) has vector relative degree $\{4,4,4,2\}$. Thus, if $C_{\mathrm{D}} = 0_{3\times3}$, then

$$\begin{bmatrix} r^{(4)}(t) \\ \ddot{\psi}(t) \end{bmatrix} = f_3(r(t), \phi(t), \theta(t), \psi(t), \omega(t), u_1(t)) + G_3^{-1}(r(t), \phi(t), \theta(t), \psi(t), u_1(t)) \begin{bmatrix} \eta_1(t) \\ u_2(t) \\ u_3(t) \\ u_4(t) \end{bmatrix},$$
$$\begin{bmatrix} r(t_0) \\ \dot{r}(t_0) \end{bmatrix} = \begin{bmatrix} r_0 \\ v_0 \end{bmatrix}, \quad \begin{bmatrix} \ddot{r}(t_0) \\ \dddot{r}(t_0) \end{bmatrix} = \begin{bmatrix} a_0 \\ j_0 \end{bmatrix}, \quad \begin{bmatrix} \psi(t_0) \\ \dot{\psi}(t_0) \end{bmatrix} = \begin{bmatrix} \psi_0 \\ \psi_{\mathrm{d},0} \end{bmatrix}, \quad t \in [t_0, \infty), \tag{50}$$

where

$$G_3^{-1}(r, \phi, \theta, \psi, u_1) \triangleq m \begin{bmatrix} \tilde{R}(\phi, \theta, \psi) & 0_{3\times1} \\ \begin{bmatrix} -\frac{I_{33}\mathrm{c}\psi\mathrm{c}\theta\mathrm{s}\phi}{u_1\mathrm{c}\phi} & -\frac{I_{33}\mathrm{c}\theta\mathrm{s}\psi\mathrm{s}\phi}{u_1\mathrm{c}\phi} & \frac{I_{33}\mathrm{s}\theta\mathrm{s}\phi}{u_1\mathrm{c}\phi} \end{bmatrix} & \frac{I_{33}\mathrm{c}\theta}{m\mathrm{c}\phi} \end{bmatrix}, \tag{51}$$

$$\tilde{R}(\phi, \theta, \psi) \triangleq \begin{bmatrix} J\tau(\mathrm{s}\phi\mathrm{s}\psi + \mathrm{c}\phi\mathrm{c}\psi\mathrm{s}\theta) & J\tau(\mathrm{c}\phi\mathrm{s}\psi\mathrm{s}\theta - \mathrm{c}\psi\mathrm{s}\phi) & J\tau\mathrm{c}\phi\mathrm{c}\theta \\ \frac{I_{11}(\mathrm{c}\phi\mathrm{s}\psi - \mathrm{c}\psi\mathrm{s}\phi\mathrm{s}\theta)}{u_1} & -\frac{I_{11}(\mathrm{c}\phi\mathrm{c}\psi + \mathrm{s}\psi\mathrm{s}\phi\mathrm{s}\theta)}{u_1} & -\frac{I_{11}\mathrm{c}\theta\mathrm{s}\phi}{u_1} \\ \frac{I_{22}\mathrm{c}\psi\mathrm{c}\theta}{u_1} & \frac{I_{22}\mathrm{c}\theta\mathrm{s}\psi}{u_1} & -\frac{I_{22}\mathrm{s}\theta}{u_1} \end{bmatrix}, \tag{52}$$

$\mathrm{c}\alpha \triangleq \cos\alpha$, $\alpha \in \mathbb{R}$, $\mathrm{s}\alpha \triangleq \sin\alpha$, and $f_3 : \mathbb{R}^3 \times \left(-\frac{\pi}{2}, \frac{\pi}{2}\right) \times \left(-\frac{\pi}{2}, \frac{\pi}{2}\right) \times [0, 2\pi) \times \mathbb{R}^3 \times \mathbb{R} \to \mathbb{R}^4$; an expression for $f_3(\cdot, \cdot, \cdot, \cdot, \cdot, \cdot)$ is omitted for brevity. It holds that

$$\det G_3(r, \phi, \theta, \psi, u_1) = \frac{u_1^2 \cos\phi}{J\tau m^3 \det(I) \cos\theta}, \quad (r, \phi, \theta, \psi, u_1) \in \mathbb{R}^3 \times \left(-\frac{\pi}{2}, \frac{\pi}{2}\right) \times \left(-\frac{\pi}{2}, \frac{\pi}{2}\right) \times [0, 2\pi) \times (0, \infty), \tag{53}$$

and, hence, $G_3(\cdot, \cdot, \cdot, \cdot, \cdot)$ is invertible if and only if $u_1 \neq 0$ since $\phi \in \left(-\frac{\pi}{2}, \frac{\pi}{2}\right)$. Furthermore, $G_3^{-1}(\cdot, \cdot, \cdot, \cdot, \cdot)$ is well-defined if and only if $u_1 \neq 0$ since $\phi \in \left(-\frac{\pi}{2}, \frac{\pi}{2}\right)$. Remarkably, if $I_{11} = I_{22} = u_1 = \tau = J = 1$, then $\tilde{R}(\cdot)$ is a rotation matrix. The hypothesis whereby $C_{\mathrm{D}} = 0_{3\times3}$ will be lifted in Section 6.5 below.

6.2 Feedback linearization relative to position and pitch angle

By proceeding as in Section 6.1, and setting $z_2(t) \triangleq [r^{\mathrm{T}}(t), \theta(t)]^{\mathrm{T}}$, $t \in [t_0, \infty)$, as a *linearizing output*, the dynamical system given by (41)–(44) and (48) has vector relative degree {4,4,4,2}, and, if $C_{\mathrm{D}} = 0_{3\times 3}$, then

$$\begin{bmatrix} r^{(4)}(t) \\ \ddot{\theta}(t) \end{bmatrix} = f_2(r(t), \phi(t), \theta(t), \psi(t), \omega(t), u_1(t)) + G_2^{-1}(r(t), \phi(t), \theta(t), \psi(t), u_1(t)) \begin{bmatrix} \eta_1(t) \\ u_2(t) \\ u_3(t) \\ u_4(t) \end{bmatrix},$$
$$\begin{bmatrix} r(t_0) \\ \dot{r}(t_0) \end{bmatrix} = \begin{bmatrix} r_0 \\ v_0 \end{bmatrix}, \begin{bmatrix} \ddot{r}(t_0) \\ \dddot{r}(t_0) \end{bmatrix} = \begin{bmatrix} a_0 \\ j_0 \end{bmatrix}, \begin{bmatrix} \theta(t_0) \\ \dot{\theta}(t_0) \end{bmatrix} = \begin{bmatrix} \theta_0 \\ \theta_{\mathrm{d},0} \end{bmatrix}, \quad t \in [t_0, \infty), \tag{54}$$

where

$$G_2^{-1}(r, \phi, \theta, \psi, u_1) \triangleq m \begin{bmatrix} \tilde{R}(\phi, \theta, \psi) & 0_{3\times 1} \\ \begin{bmatrix} \dfrac{I_{33}\mathrm{c}\phi\mathrm{c}\psi\mathrm{c}\theta}{u_1\mathrm{s}\phi} & \dfrac{I_{33}\mathrm{c}\phi\mathrm{c}\theta\mathrm{s}\psi}{u_1\mathrm{s}\phi} & \dfrac{-I_{33}\mathrm{c}\phi\mathrm{s}\theta}{u_1\mathrm{s}\phi} \end{bmatrix} & \dfrac{-I_{33}}{m\mathrm{s}\phi} \end{bmatrix}, \tag{55}$$

$f_2 : \mathbb{R}^3 \times \left(-\frac{\pi}{2}, \frac{\pi}{2}\right) \times \left(-\frac{\pi}{2}, \frac{\pi}{2}\right) \times [0, 2\pi) \times \mathbb{R}^3 \times \mathbb{R} \to \mathbb{R}^4$; an expression for $f_2(\cdot, \cdot, \cdot, \cdot, \cdot, \cdot)$ is omitted for brevity. It holds that

$$\det G_2(r, \phi, \theta, \psi, u_1) = \frac{-u_1^2 \sin\phi}{J\tau m^3 \det I}, \quad (r, \phi, \theta, \psi, u_1) \in \mathbb{R}^3 \times \left(-\frac{\pi}{2}, \frac{\pi}{2}\right) \times \left(-\frac{\pi}{2}, \frac{\pi}{2}\right) \times [0, 2\pi) \times (0, \infty), \tag{56}$$

and, hence, $G_2(\cdot, \cdot, \cdot, \cdot, \cdot)$ is invertible if and only if $u_1 \neq 0$ and $\phi \neq 0$ since $\phi \in \left(-\frac{\pi}{2}, \frac{\pi}{2}\right)$. Furthermore, $G_2^{-1}(\cdot, \cdot, \cdot, \cdot, \cdot)$ is well-defined if and only if $u_1 \neq 0$ and $\phi \neq 0$.

6.3 Feedback linearization relative to position and roll angle

Setting $z_1(t) \triangleq [r^{\mathrm{T}}(t), \phi(t)]^{\mathrm{T}}$, $t \in [0, \infty)$, as a *linearizing output*, the dynamical system given by (41)–(44) and (48) has vector relative degree {4,4,4,2}, and, if $C_{\mathrm{D}} = 0_{3\times 3}$, then

$$\begin{bmatrix} r^{(4)}(t) \\ \ddot{\phi}(t) \end{bmatrix} = f_1(r(t), \phi(t), \theta(t), \psi(t), \omega(t), u_1(t)) + G_1^{-1}(r(t), \phi(t), \theta(t), \psi(t), u_1(t)) \begin{bmatrix} \eta_1(t) \\ u_2(t) \\ u_3(t) \\ u_4(t) \end{bmatrix},$$
$$\begin{bmatrix} r(t_0) \\ \dot{r}(t_0) \end{bmatrix} = \begin{bmatrix} r_0 \\ v_0 \end{bmatrix}, \begin{bmatrix} \ddot{r}(t_0) \\ \dddot{r}(t_0) \end{bmatrix} = \begin{bmatrix} a_0 \\ j_0 \end{bmatrix}, \begin{bmatrix} \phi(t_0) \\ \dot{\phi}(t_0) \end{bmatrix} = \begin{bmatrix} \phi_0 \\ \phi_{d,0} \end{bmatrix}, \quad t \in [t_0, \infty), \tag{57}$$

where

$$G_1^{-1}(r, \phi, \theta, \psi, u_1) \triangleq m \begin{bmatrix} \tilde{R}(\phi, \theta, \psi) & 0_{3\times 1} \\ \begin{bmatrix} \frac{-I_{33} c\theta s\psi}{u_1 s\theta} & \frac{-I_{33} c\psi c\theta}{u_1 s\theta} & \frac{I_{33} s\phi}{u_1 c\phi s\theta} \end{bmatrix} & \frac{I_{33} c\theta}{m c\phi s\theta} \end{bmatrix}, \tag{58}$$

$f_1 : \mathbb{R}^3 \times \left(-\frac{\pi}{2}, \frac{\pi}{2}\right) \times \left(-\frac{\pi}{2}, \frac{\pi}{2}\right) \times [0, 2\pi) \times \mathbb{R}^3 \times \mathbb{R} \to \mathbb{R}^4$; an expression for $f_1(\cdot, \cdot, \cdot, \cdot, \cdot, \cdot)$ is omitted for brevity. It holds that

$$\det G_1(r, \phi, \theta, \psi, u_1) = \frac{u_1^2 \cos\phi \tan\theta}{J\tau m^3 \det I}, \quad (r, \phi, \theta, \psi, u_1) \in \mathbb{R}^3 \times \left(-\frac{\pi}{2}, \frac{\pi}{2}\right) \times \left(-\frac{\pi}{2}, \frac{\pi}{2}\right) \times [0, 2\pi) \times (0, \infty), \tag{59}$$

and hence, $G_1(\cdot, \cdot, \cdot, \cdot, \cdot)$ is invertible if and only if $u_1 \neq 0$ and $\theta \neq 0$ since $\phi \in \left(-\frac{\pi}{2}, \frac{\pi}{2}\right)$. Furthermore, $G_1^{-1}(\cdot, \cdot, \cdot, \cdot, \cdot)$ is well-defined if and only if $u_1 \neq 0$ and $\theta \neq 0$.

6.4 Feedback linearization with MRAC augmentation

In light of the results in Sections 6.1–6.3, let

$$\zeta_\sigma(r, \phi, \theta, \psi, \omega, u_1, \lambda_\sigma) \triangleq G_\sigma(r, \phi, \theta, \psi, \omega, u_1)\left(-f_\sigma(r, \phi, \theta, \psi, \omega, u_1) + \begin{bmatrix} A_{r,0} r + A_{r,1}\dot{r} + A_{r,2}\ddot{r} + A_{r,3}\dddot{r} \\ A_{y,\sigma,0} y_\sigma + A_{y,\sigma,1}\dot{y}_\sigma \end{bmatrix} + \begin{bmatrix} B_r \\ B_{y,\sigma} \end{bmatrix} \lambda_\sigma\right),$$
$$(r, \phi, \theta, \psi, \omega, u_1, \lambda_\sigma) \in \mathbb{R}^3 \times \left(-\frac{\pi}{2}, \frac{\pi}{2}\right) \times \left(-\frac{\pi}{2}, \frac{\pi}{2}\right) \times [0, 2\pi) \times \mathbb{R}^3 \times \mathbb{R} \times \mathbb{R}^4, \tag{60}$$

denote the *baseline feedback-linearizing control input*, where $\sigma \in \{1,2,3\}$, $y_1 = \phi$, $y_2 = \theta$, $y_3 = \psi$,

$$\tilde{A}_r \triangleq \begin{bmatrix} 0_{3\times 3} & \mathbf{1}_3 & 0_{3\times 3} & 0_{3\times 3} \\ 0_{3\times 3} & 0_{3\times 3} & \mathbf{1}_3 & 0_{3\times 3} \\ 0_{3\times 3} & 0_{3\times 3} & 0_{3\times 3} & \mathbf{1}_3 \\ A_{r,0} & A_{r,1} & A_{r,2} & A_{r,3} \end{bmatrix} \in \mathbb{R}^{12\times 12}, \quad \tilde{A}_{y,\sigma} \triangleq \begin{bmatrix} 0 & 1 \\ A_{y,\sigma,0} & A_{y,\sigma,1} \end{bmatrix} \in \mathbb{R}^{2\times 2}, \tag{61}$$

are Hurwitz,

$$\tilde{B}_r \triangleq \begin{bmatrix} 0_{9\times 4} \\ B_r \end{bmatrix} \in \mathbb{R}^{12\times 4}, \quad \tilde{B}_{y,\sigma} \triangleq \begin{bmatrix} 0_{1\times 4} \\ B_{y,\sigma} \end{bmatrix} \in \mathbb{R}^{2\times 4}, \tag{62}$$

and the pairs $(\tilde{A}_r, \tilde{B}_r)$ and $(\tilde{A}_{y,\sigma}, \tilde{B}_{y,\sigma})$ are controllable. If

$$[\eta_1(t), u_2(t), u_3(t), u_4(t)]^{\mathrm{T}} = \zeta_\sigma(t), \qquad t \in [t_0, \infty), \tag{63}$$

for some $\sigma \in \{1,2,3\}$, where $\zeta_\sigma(t)$ denotes $\zeta_\sigma(r(t), \phi(t), \theta(t), \psi(t), \omega(t), u_1(t), \lambda_\sigma(t))$ for brevity, then the UAV's equations of motion (41)–(44) are output-feedback-linearized and

$$\dot{\chi}_\sigma(t) = A_\sigma \chi_\sigma(t) + B_\sigma \lambda_\sigma(t), \qquad \chi_\sigma(t_0) = \chi_{\sigma,0}, \qquad t \geq t_0, \tag{64}$$

where $\chi_\sigma(t) \triangleq [r^{\mathrm{T}}(t), \dot{r}^{\mathrm{T}}(t), \ddot{r}^{\mathrm{T}}(t), \dddot{r}^{\mathrm{T}}(t), y_\sigma(t), \dot{y}_\sigma(t)]^{\mathrm{T}} \in \mathbb{R}^{14}$, $A_\sigma \triangleq \mathrm{blockdiag}(\tilde{A}_r, \tilde{A}_{y,\sigma})$ $\in \mathbb{R}^{14\times 14}$, $B_\sigma \triangleq \left[\tilde{B}_r^{\mathrm{T}}, \tilde{B}_{y,\sigma}^{\mathrm{T}}\right]^{\mathrm{T}} \in \mathbb{R}^{14\times 4}$, and the initial condition $\chi_{\sigma,0} \in \mathbb{R}^{14}$ deduced from (50), (54), and (57).

Fixed $\sigma \in \{1,2,3\}$, to account for the fact that, in general, $C_{\mathrm{D}} \neq 0_{3\times 3}$, we generalize (64) and consider the plant model

$$\begin{aligned} \dot{\chi}_\sigma(t) &= A_\sigma \chi_\sigma(t) + B_\sigma \Lambda_\sigma \left[\lambda_\sigma(t) + \Theta_\sigma^{\mathrm{T}} \Phi_\sigma(t, \chi_\sigma(t))\right] + \begin{bmatrix} \xi_\sigma(t) \\ 0_2 \end{bmatrix}, \\ \chi_\sigma(t_0) &= \chi_{\sigma,0}, \qquad t \geq t_0, \end{aligned} \tag{65}$$

where $\Lambda_\sigma \in \mathbb{R}^{4\times 4}$ is diagonal, positive-definite, and unknown. By setting $\Lambda_\sigma = \mathrm{diag}(m^{-1}, m^{-1}, m^{-1}, I_{\sigma\sigma}^{-1})$, $\sigma \in \{1,2,3\}$, this matrix can be employed to account for uncertainties in the UAV's mass and moment of inertia corresponding to the selected linearizing output signal z_σ. The unmatched uncertainty

$$\xi_\sigma(t) \triangleq \left[0_3^{\mathrm{T}}, \xi_{\mathrm{drag},\sigma}^{\mathrm{T}}(t, v(t)), 0_6^{\mathrm{T}}\right]^{\mathrm{T}} \in \mathbb{R}^{12}, \tag{66}$$

where

$$\begin{aligned} \xi_{\mathrm{drag},\sigma}(t, v) &\triangleq -\frac{1}{2m} \rho S R^{\mathrm{T}}(\phi(t), \theta(t), \psi(t)) C_{\mathrm{D}} \|v\| v, \\ (t, v) &\in [t_0, \infty) \times \left(\mathbb{R}^3 \backslash \{0_3\}\right), \end{aligned} \tag{67}$$

captures the effect of aerodynamic forces, which are not explicitly accounted for in the feedback-linearizing control law (60). The regressor vector $\Phi_\sigma : [t_0, \infty) \times \mathbb{R}^{14} \to \mathbb{R}^{N_\sigma}$ includes the baseline controller and matched parametric uncertainties not accounted for in the feedback-linearization process. To capture uncertainties in the feedback-linearized plant dynamics, such as uncertainties in the location of the UAV's center of mass, such a regressor vector can be constructed to be an explicit function of the UAV's translational and angular position as well as of the UAV's rotational position and velocity, thus linking explicitly (65) to (41)–(44). Explicit expressions of $\Phi_\sigma(\cdot,\cdot)$, $\sigma \in \{1,2,3\}$, will be presented in future works.

Having reduced the feedback-linearized equations of motion of the UAV to the classical form or MRAC, we can compute the *virtual control input* $\lambda_\sigma(\cdot)$ so that the feedback-linearized plant trajectory $\chi_\sigma(\cdot)$ follows the reference trajectory $\chi_{\text{ref},\sigma}$: $[t_0,\infty) \to \mathbb{R}^{14}$ such that

$$\dot{\chi}_{\text{ref},\sigma}(t) = A_{\text{ref},\sigma}\chi_{\text{ref},\sigma} + B_{\text{ref},\sigma}r_\sigma(t), \qquad \chi_{\text{ref},\sigma}(t_0) = \chi_{\text{ref},\sigma,0}, \qquad t \geq t_0, \tag{68}$$

where $A_{\text{ref},\sigma} \in \mathbb{R}^{14\times14}$ is Hurwitz, $B_{\text{ref},\sigma} \in \mathbb{R}^{14\times4}$ is such that the pair $(A_{\text{ref},\sigma}, B_{\text{ref},\sigma})$ is controllable, and $r_\sigma : [t_0,\infty) \to \mathbb{R}^4$ denotes the user-defined *reference command input*. Fixing $\sigma \in \{1,2,3\}$, this task can be attained by employing a robust MRAC system or any other nonlinear robust control technique, such as sliding mode or any of its higher-order variations.

Since A_σ is user-defined and Hurwitz, $\sigma \in \{1,2,3\}$, and both B_σ and $B_{\text{ref},\sigma}$ are user-defined, it is possible to set $A_{\text{ref},\sigma} = A_\sigma$ and $B_{\text{ref},\sigma} = B_\sigma$. Furthermore, $r(\cdot)$ can be designed so that $\chi_{\text{ref},\sigma}(\cdot)$ follows the user-defined signal $\chi_{\text{user},\sigma} : [t_0,\infty) \to \mathbb{R}^{14}$, whose first 12 components capture the desired position, velocity, acceleration, and jerk, and whose last 2 components capture the desired trajectory for the UAV's measured angle and angular rate.

6.5 Hybrid MRAC and feedback linearization

If $\sigma : [t_0,\infty) \to \{1,2,3\}$ is a function of time, then the control system presented in Section 4 can be applied to compute the virtual control input $\lambda_{\sigma(\cdot)}(\cdot)$. Indeed, (65) is in the same form as the continuous-time plant dynamics given by (8) with $\tilde{\Theta}_{\sigma(t)} = 0$ and $\Sigma = \{1,2,3\}$. Similarly, (68) is in the same form as the continuous-time reference model dynamics given by (10).

The sets of resetting events $\{\mathcal{S}_\sigma\}_{\sigma\in\Sigma}$, which characterize the switching among the lineatizing outputs $z_{\sigma(t)}(t)$, $t \geq t_0$, are provided by Algorithm 1. This algorithm assumes that the user provides a four time continuously differentiable desired trajectory for the UAV's position and a twice continuously differentiable desired trajectory for the yaw, pitch, and roll angles. The user-defined trajectories for the roll and pitch angles are such that $|\phi_{\text{user}}(t)| \in (\phi_{\min}, \phi_{\max})$, $t \geq t_0$, and $|\theta_{\text{user}}(t)| \in (\theta_{\min}, \theta_{\max})$, where $0 < \phi_{\min} < \phi_{\max}$ and $0 < \theta_{\min} < \theta_{\max}$.

Algorithm 1: Algorithm for multi-output feedback linearization.

1: $t^* \leftarrow t_0$ ▷ Initialize the last switching time variable
2: for $t \geq t_0$ **do**
3: $T_i(t) \leftarrow \text{sat}(T_i(t), T_{i,\min}, T_{i,\max})$, $i \in \{1,\dots,4\}$ ▷ Enforce saturation constraints on thrust force $T_i(t)$
4: **if** $\sigma(t) = 3t - t^* \geq \Delta T_{\min}$ **then**
5: **if** $|\phi(t)| \geq \phi_{\max}$ OR $|\theta(t)| \geq \theta_{\max}$ **then**
6: $\sigma(t) \leftarrow \arg\max\{|\phi(t)| - \phi_{\max}, |\theta(t)| - \theta_{\max}\}$
7: $t^* \leftarrow t$
8: **end if**

9: **else if** $\sigma(t) = 2$ & ($|\phi(t)| \leq \phi_{\min}$ OR $|\phi(t)| \geq \phi_{\max}$)&$t - t^* \geq \Delta T_{\min}$ then
▷ Enforce constraints on $G_2(\cdot)$
10: $\sigma(t) \leftarrow \arg\max_{\sigma \in \{1,3\}} \left(\|e_{z,\sigma(t)}(t)\| - \varepsilon_\sigma\right)$
11: $t^* \leftarrow t$
12: **else if** $\sigma(t) = 1$ ($|\theta(t)| \leq \theta_{\min}$ OR $|\theta(t)| \geq \theta_{\max}$)$t - t^* \geq \Delta T_{\min}$ then
▷ Enforce the constraints on $G_1(\cdot)$
13: $\sigma(t) \leftarrow \arg\max_{\sigma \in \{2,3\}} \left(\|e_{z,\sigma(t)}(t)\| - \varepsilon_\sigma\right)$
14: $t^* \leftarrow t$
15: **end if**
16: **if** $\|e_{z,\sigma(t)}(t)\| > \varepsilon_\sigma t - t^* \geq \Delta T_{\min}$ then ▷ If any of the tracking errors is too large and enough time has passed since the last switching
17: $\sigma(t) \leftarrow \arg\max_{\sigma \in \{1,2,3\}} \left(\|e_{z,\sigma(t)}(t)\| - \varepsilon_\sigma\right)$
18: $t^* \leftarrow t$
19: **end if**
20: **end for**

To present Algorithm 1, let the user-defined variable $\Delta T_{\min} > 0$ denote the *dwell time* of the plant model, that is, the minimum time between two consecutive switching of the index $\sigma(\cdot)$. Furthermore, for each $\sigma \in \{1, 2, 3\}$, let $\varepsilon_\sigma > 0$ denote the user-defined tolerance on the *output signal tracking error*

$$e_{z,\sigma(t)}(t) \triangleq C\left(\chi_{\sigma(t)}(t) - \chi_{\mathrm{ref},\sigma(t)}(t)\right), \qquad t \geq t_0, \tag{69}$$

where $C \triangleq \begin{bmatrix} \mathbf{1}_3 & 0_{3\times3} & 0_{3\times6} & 0_{3\times2} \\ 0_{3\times3} & \mathbf{1}_3 & 0_{3\times6} & 0_{3\times2} \\ 0_{2\times3} & 0_{2\times3} & 0_{2\times6} & \mathbf{1}_2 \end{bmatrix} \in \mathbb{R}^{8\times14}$. Additionally, let $T_{i,\min} > 0$, $i \in \{1,2,3,4\}$, and $T_{i,\max} > T_{i,\min}$ denote the minimum and maximum allowed thrust for the ith motor, respectively. Finally, let

$$\begin{aligned} &\mathrm{sat}(\alpha, \alpha_{\min}, \alpha_{\max}) \triangleq \min\{\alpha_{\max}, \max\{\alpha, \alpha_{\min}\}\}, \\ &\quad (\alpha, \alpha_{\min}, \alpha_{\max}) \in \mathbb{R} \times \mathbb{R} \times \mathbb{R}, \end{aligned} \tag{70}$$

denote the *saturation function*, where $\alpha_{\min} < \alpha_{\max}$.

7. Numerical simulation results

In this section, we illustrate the applicability of the proposed results by means of a numerical simulation. In this simulation, the UAV is tasked with ascending and moving at a constant velocity along the X-axis of the inertial reference frame for $t \in [0, 5]$ s, hovering for $t \in [5, 10]$ s, following an upward spiral trajectory for $t \in [10, 30]$ s, descending along the same spiral for $t \in [30, 50]$ s, translating along the bissetriz of the horizontal plane at a constant velocity for $t \in [50, 60]$ s, and hovering until the end of the mission. Furthermore, the user requires that the UAV's yaw, pitch, and roll angles follow predefined trajectories within a margin of 5 degrees at all times. The user-defined yaw and roll angles are constant at all times, and the user-defined pitch angle is linearly increasing in the ascending and descending phases of the spiral trajectory and constant everywhere else; for details, see **Figure 1**. It is worthwhile to remark that this reference attitude poses a significant challenge. Indeed, as discussed in Sections 6.2 and 6.3, if $\sigma = 2$, that is, if the feedback linearizing output comprises the UAV's position and pitch angle, then the roll angle can not be equal to zero, Similarly, if $\sigma = 1$, that is, if the feedback linearizing output comprises the UAV's position and roll angle, then the pitch angle can not be set to zero. Furthermore, to follow the reference spiral trajectory imposed by the user, the roll and pitch angles must vary sinusoidally.

Figure 2 shows the thrust force and the time derivative of the thrust force needed by the UAV to follow the user-defined trajectory. Both $u_1(t)$, $t \geq 0$, and $\dot{u}_1(t)$ show

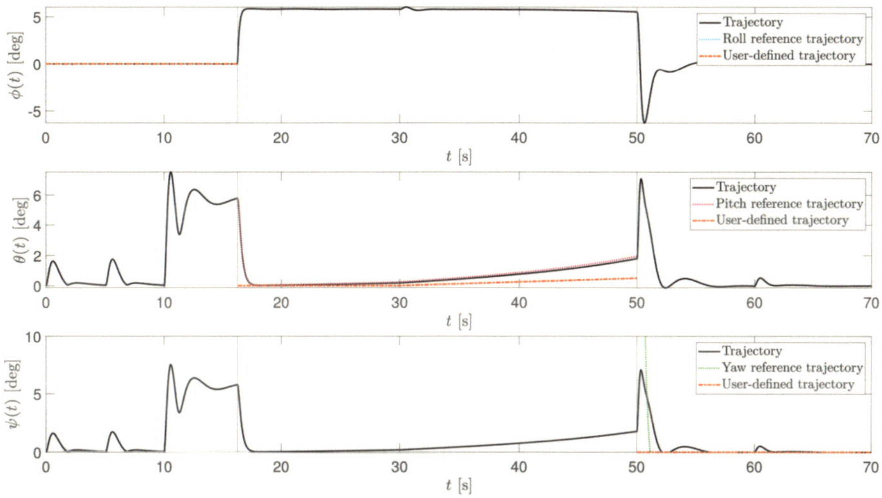

Figure 1.
Euler angles capturing the attitude of the UAV. At $t = 0$ s, the feedback linearizing output is set as $\sigma = 1$. At $t = 16.0953$ s, shortly after the UAV is tasked with hovering, applying Algorithm 1, the control system switches feedback linearizing output to $\sigma = 2$. Finally, at $t = 50.0362$ s, before the UAV is tasked with moving sideways in the horizontal plane, applying Algorithm 1, the control system sets $\sigma = 3$. In this stage, after a brief transient, the yaw angle closely follows its reference trajectory.

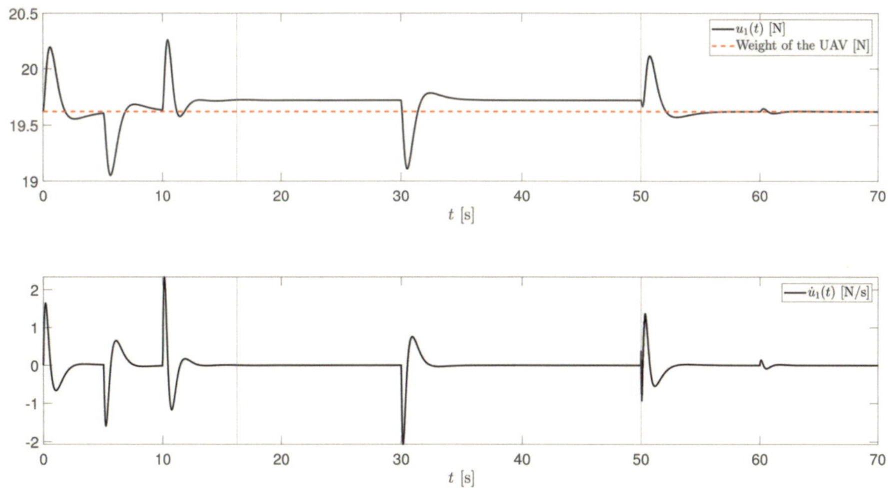

Figure 2.
Total thrust and time derivative of the total thrust. The total thrust and its derivative lie within bounds that are typical for commercial-off-the-shelf motors of Class 1 quadcopter UAVs.

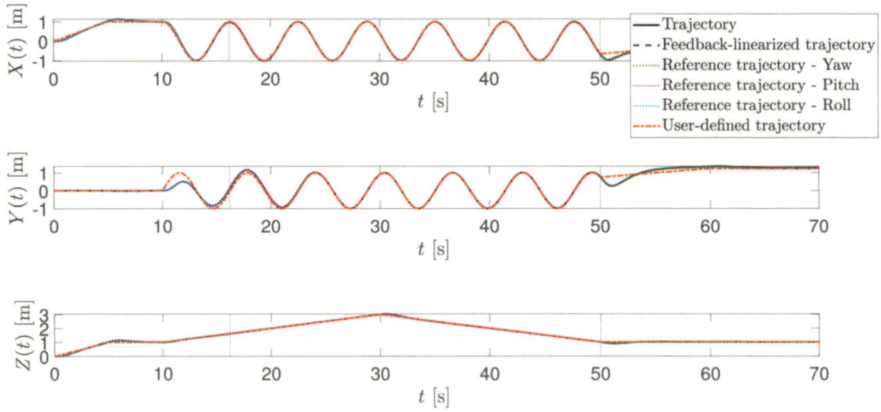

Figure 3.
Trajectory of the center of mass of the UAV as a function of time. In all modes, the vehicle's trajectory closely follows the user-defined trajectory despite uncertainties and the drag force.

profiles that are compatible with the performances of commercial-off-the-shelf electric motors for Class 1 quadcopter UAVs.

In this simulation, the UAV's mass is $m = 2$ kg and its central matrix of inertia is given by $I = \text{diag}(0.010, 0.010, 0.015)$ kg m^2. The matrix of aerodynamic coefficient is set equal to $C_D = 0.001 \cdot \mathbf{1}_3$. The estimated mass is 2.2 kg and the estimated matrix of inertia is given by diag(0.020,0.015,0.025) kg m^2. The adaptive rate matrix is set as $\Gamma_\sigma = 9 \cdot 10^2 \cdot \mathbf{1}_{22}$ for all $\sigma \in \{1,2,3\}$. We set

$$B_r = \begin{bmatrix} \mathbf{1}_3 & 0_3 \\ [1 \quad 1 \quad 1] & 1 \end{bmatrix}, \quad B_{y,\sigma} = [1 \quad 1 \quad 1], \tag{71}$$

and both $\tilde{A}_r$ and $\tilde{A}_{y,\sigma}$ were designed through the pole placement method, imposing eigenvalues $\{-4.7, -5, -2.5, -2.6, -2.9, -8, -1.21, -2.3, -1.6, -1.8, -1.5, -1.6\}$ for the translational dynamics, eigenvalues $\{-4, -7\}$ for $\sigma = 3$ and eigenvalues $\{-4, -8\}$ for $\sigma \in \{1, 2\}$. The σ-modification of the MRAC, that is, the adaptive law (22) is employed with $\gamma_\sigma = 0.01$ for all $\sigma \in \{1,2,3\}$.

Figure 3 shows the UAV trajectory as a function of time. It is apparent how the UAV closely follows the reference trajectory at all times. **Figure 1** shows the UAV attitude by means of the yaw, pitch, and roll angles. The reference angle as well as the user-defined angle are shown only for those stages in which the mode is active. At $t = 0$ s, the feedback linearizing output is set as $\sigma = 1$. At $t = 16.0953$ s, shortly after the UAV is tasked with hovering, applying Algorithm 1, the control system switches feedback linearizing output to $\sigma = 2$. Finally, at $t = 50.0362$ s, before the UAV is tasked with moving sideways in the horizontal plane, applying Algorithm 1, the control system sets $\sigma = 3$. In this stage, after a brief transient, the yaw angle closely follows its reference trajectory. Numerical evidence show that, without the proposed hybrid system, this maneuver would not be possible by setting $\sigma(t) \equiv 1$ or $\sigma(t) \equiv 2$ for all $t \geq 0$, that is, without the proposed control system.

DOI: http://dx.doi.org/10.5772/intechopen.1004814

8. Conclusion

This chapter presented the first robust MRAC system applicable to time-varying, hybrid plant models affected by parametric, matched, and unmatched uncertainties in the continuous-time dynamics as well as uncertainties in the discrete-time dynamics. These results have been applied to the problem of controlling the feedback-linearized dynamics of a quadcopter UAV and tasking the vehicle to follow both a user-defined trajectory and a user-defined attitude. This result is unprecedented because, due to the UAV's underactuation, existing works on the control of quadcopters allow regulating arbitrarily only four of its six degrees of freedom. The proposed approach, instead, allows the user to impose reference trajectories for each of the UAV's six degrees of freedom. Future work directions concern the extension of the proposed approach from a specific application, namely quadcopter UAVs, to generic plant models.

Future work directions involve further extensions of the proposed hybrid MRAC framework for the control of output-feedback linearized systems to cases wherein the feedback-linearizing output is affected by noise. Additional work directions include problems wherein the feedback-linearizing output is not readily available for measurement but needs to be deduced from the measured output.

Acknowledgements

This research was in part performed with the support of NSF through the Grant no. 2137159 and the US Army Research Lab under Grant no. W911QX2320001.

Conflict of interest

The authors declare no conflict of interest.

Abbreviations

MRAC	model reference adaptive control
UAV	unmanned aerial vehicle

Author details

Giri M. Kumar[1†], Mattia Gramuglia[†] and Andrea L'Afflitto[2*]

1 Department of Mechanical Engineering, Virginia Tech, Blacksburg, VA, USA

2 Grado Department of Industrial and Systems Engineering, Virginia Tech, Blacksburg, VA, USA

*Address all correspondence to: a.lafflitto@vt.edu

† These authors contributed equally.

References

[1] Cortes J. Discontinuous dynamical systems. IEEE Control Systems Magazine. 2008;**28**(3):36-73

[2] Anderson RB, Marshall JA, L'Afflitto A, Dotterweich JM. Model reference adaptive control of switched dynamical systems with applications to aerial robotics. Journal of Intelligent & Robotic Systems. 2020; **100**(3):1265-1281

[3] Lin H, Antsaklis P. Hybrid Dynamical Systems: Fundamentals and Methods, Ser. Advanced Textbooks in Control and Signal Processing. London, UK: Springer; 2021

[4] L'Afflitto A. AfflittoModel reference adaptive control for nonlinear time-varying hybrid dynamical systems. International Journal of Adaptive Control and Signal Processing. 2023; **37**(8):2162-2183

[5] Narendra K, Annaswamy A. A new adaptive law for robust adaptation without persistent excitation. IEEE Transactions on Automatic Control. 1987;**32**(2):134-145

[6] Ioannou P, Kokotovic P. Adaptive Systems with Reduced Models. New York, NY: Springer; 1983

[7] Ioannou P, Fidan B. Adaptive Control Tutorial. Philadelphia, PA: Society for Industrial and Applied Mathematics; 2006

[8] Isidori A. Nonlinear Control Systems. New York, NY: Springer; 1995

[9] Malo Tamayo AJ, Villaseñor Rios CA, Ibarra Zannatha JM, Orozco Soto SM. Quadrotor Input-Output Linearization and Cascade Control. IFAC-PapersOnLine. 2018;**51**(13):437-442. ISSN: 2405-8963. DOI: 10.1016/j.ifacol.2018.07.317

[10] Lotufo MA, Colangelo L, Novara C. Control design for uav quadrotors via embedded model control. Transactions on Control Systems Technology. 2020; **28**(5):1741-1756

[11] Daga GD, Thosar AG. Feedback linearization of quadrotor. In: Shrivastava V, Bansal JC, Panigrahi BK, editors. Power Engineering and Intelligent Systems. Singapore: Springer; 2024. pp. 137-151

[12] PX4 Website. Available from: https://px4.io/ [Accessed: February 12, 2024]

[13] Ardupilot Website. Available from: https://ardupilot.org/ [Accessed: February 12, 2024]

[14] Kreyszig E. Introductory Functional Analysis with Applications. New York, NY: Wiley; 1989

[15] Bernstein DS. Matrix Mathematics: Theory, Facts, and Formulas. 2nd ed. Princeton, NJ: Princeton University Press; 2009

[16] Haddad WM, Chellaboina VS, Nersesov SG. Impulsive and Hybrid Dynamical Systems: Stability, Dissipativity, and Control. Princeton, NJ: Princeton; 2014

[17] Sanfelice RG, Goebel R, Teel AR. Generalized solutions to hybrid dynamical systems. Control, Optimisation and Calculus of Variations. 2008;**14**(4):699-724

[18] Bardi M, Capuzzo-Dolcetta I. Optimal Control and Viscosity Solutions

of Hamilton-Jacobi-Bellman Equations. Boston, MA: Birkhäuser; 2008

[19] Clarke FH. Optimization and Nonsmooth Analysis. Philadelphia, PA: Society of Industrial and Applied Mathematics; 1989

[20] Fischer N, Kamalapurkar R, Dixon WE. LaSalle-Yoshizawa corollaries for nonsmooth systems. IEEE Transactions on Automatic Control. 2013;**58**(9):2333-2338

[21] Goebel R, Sanfelice R, Teel A. Hybrid Dynamical Systems: Modeling, Stability, and Robustness. Princeton University Press; 2012. Available form: http://www.jstor.org/stable/j.ctt7s02z. ISBN: 9780691153896

[22] Khalil HK. Nonlinear Systems. Princeton, NJ: Prentice Hall; 2002

[23] Pomet JB, Praly L. Adaptive nonlinear regulation: estimation from the Lyapunov equation. IEEE Transactions on Automatic Control. 1992;**37**(6):729-740

[24] Lavretsky E, Wise K. Robust and Adaptive Control: With Aerospace Applications. London, UK: Springer; 2012

[25] Ioannou P, Sun J. Robust Adaptive Control, Ser. Dover Books on Electrical Engineering Series. Upper Saddle River, NJ: Dover Publications; 2012

[26] L'Afflitto A, Anderson RB, Mohammadi K. An introduction to nonlinear robust control for unmanned quadrotor aircraft. IEEE Control Systems Magazine. 2018;**38**(3):102-121

[27] L'Afflitto A. A Mathematical Perspective on Flight Dynamics and Control. London, UK: Springer; 2017

Chapter 5

Decentralized Robust Direct MRAC for the Attitude of a Quadrotor UAV

Francisco Jurado

Abstract

In this chapter, in order to prevent the parameter drift in the adaptive control of the attitude for a quadrotor unmanned aerial vehicle (UAV), a decentralized robust direct model reference adaptive controller (MRAC) is proposed. The dynamic model for the quadrotor UAV referred to as the rotational system is parameterized in a decentralized way, that is, for each one of the angles of motion, in terms of a matched uncertainty by the control input and considering the presence of bounded perturbations. A modification on the update law to enforce robustness is implemented in the design of the proposal. A linear PID controller is considered for the control of the translational dynamics. The performance of the decentralized robust direct MRACs is evaluated *via* numerical simulation results.

Keywords: adaptive control, decentralized control, parametric drift, model reference control, parametric identification, robust control, unmanned aerial vehicles

1. Introduction

In some applications, the output is influenced not only by the input, usually referred to as the controlled input, but also by other uncontrolled inputs. Such inputs create perturbations that may influence the system in unpredictable ways. In general, external perturbations are time-varying functions that may influence the output persistently and may be present for a limited time. In some cases, perturbations may exhibit known time-varying characteristics; namely, these can be of unknown amplitude but with known frequency.

In this work, modification to the standard (parameter estimator) learning algorithm is considered in order to improve performance and ensure stability in the presence of modeling errors and external perturbations into the control of the attitude for a quadrotor UAV. This modification leads to what is known as robust learning algorithms. A learning algorithm that preserves stability properties under modeling errors and external perturbations, within some specifications of design, is then referred to as a robust learning algorithm [1–3]. From the linear systems adaptive control literature, it is well known that in the presence of measurement noise, unmodeled dynamics, time variations, and external perturbations, the conventional *update (adaptive) laws* show *parameter drift*, this latter a phenomenon in which the parameter estimates drift, possibly to infinity, from their ideal values. Parameter drift occurs as a result of the online parameter estimator when a matching condition does not exist any value for the parameters when attempting an exact match for a function. There are two alternatives to anticipate parameter drift. The first one consists of indirectly not performing parameter adaptation when the training error is too small. The dead-zone modification [4] technique falls in this category. In the last one, the online parameter estimator is modified in such a way that the parameter estimates are directly restricted from the drift to infinity. The σ-modification [5], also known as leakage factor [1, 6], e-modification [7, 8], and projection algorithms [1, 6, 9–11] pertain to this last category.

The design of stable adaptive controllers is the essential objective when trying with asymptotic model following for systems free of perturbations. Perturbations such as unmodeled dynamics and time-varying parameters could result in unbounded solutions. The exact model following is no longer possible for systems under the effects of the presence of perturbations [12]. To this last eventuality, the boundedness of all the signals from the closed-loop system is the main goal where the controller design task must satisfy certain performance conditions. So, robust adaptive control approaches are required to assure the boundedness of the signals in the system. In this context, robustness implies that the adaptive system performs basically in the same manner even when external perturbations are present. In order to reach the desired goal two approaches can be adopted. In the first approach, the updated law is modified properly. In the second approach, the conventional adaptive law does not suffer from any modification. Instead, to ensure the boundedness of the signals in the adaptive system, a reference input that satisfies the well-known *persistence of excitation* property, that is, a sufficiently rich input signal, is invoked.

In the presence of measurement noise, if the input signal is not sufficiently rich, the parameters just not only do not converge to their nominal values, these may also drift. Parameter drift can be avoided by using sufficiently rich inputs. When such an option is not possible, update law modifications, namely dead-zone modification, σ-modification, and e-modification, can be considered.

An adaptive controller is defined to be robust if it guarantees signal boundedness in the presence of reasonable classes of unmodeled dynamics and bounded perturbations as well as performance error bounds that are of the order of the modeling error [1, 2, 4, 5, 13].

In this chapter, a decentralized robust direct MRAC for the attitude of a quadrotor UAV is proposed. The model for the quadrotor UAV is given in terms of a decentralized uncertain model where the uncertainties enter into the subsystem dynamics *via* control channels; that is, the uncertainties are matched into the subsystem. This latter in order to add an extra realism to the ideal system. Also, an adaptive control law is designed for each one of the subsystems that constitute the entire system. Moreover, in order to face the presence of modeling uncertainties and external perturbations, a modification to the updated law is implemented to preventing parameter drift. The proposal is validated *via* numerical simulation results.

2. Dynamic model of a quadrotor

In this work a miniature four-rotor helicopter is considered, having two of these rotors rotating clockwise, and two rotating counterclockwise. Each rotor consists of an electric DC motor, a drive gear, and a rotor blade. Forward motion is accomplished by increasing the speed of the rear rotor while simultaneously reducing the forward rotor by the same amount. Backward, leftward, and rightward motion can be accomplished in the same way. Yaw motion can be performed by accelerating the two clockwise turning rotors while decelerating the counterclockwise ones.

The governing equations of motion that describe the attitude and position of a quadrotor are those from a rigid body with six degrees of freedom [14]. These latter can be decoupled into kinematic and dynamic equations [15]. The dynamic equations

$$F_{\text{ext}} = m_q \dot{V}_b + \omega \times m_q V_b, \tag{1}$$

$$T_{\text{ext}} = J\dot{\omega} + \omega \times J\omega, \tag{2}$$

can be obtained around the center of mass where m_q denotes the quadrotor mass, V_b is the velocity in the body frame, ω is the angular rate of the quadrotor, $J = \text{diag}\left\{J_x, J_y, J_z\right\}$ is the inertia matrix, the external force $F_{\text{ext}} \in \mathbb{R}^3$ takes into account the quadrotor weight, the total thrust, and the aerodynamic force, whereas the external torque $T_{\text{ext}} \in \mathbb{R}^3$ considers the difference of thrust and torque exerted by the two pairs of rotors as well as the aerodynamic moment vector. From a Z-Y-X sequence, consider the rotation matrix

$$\mathscr{R}(\psi, \theta, \phi) = \begin{pmatrix} \text{c}\theta\text{c}\psi & \text{c}\psi\text{s}\theta\text{s}\phi - \text{c}\phi\text{s}\psi & \text{c}\phi\text{c}\psi\text{s}\theta + \text{s}\phi\text{s}\psi \\ \text{c}\theta\text{s}\psi & \text{s}\theta\text{s}\phi\text{s}\psi + \text{c}\phi\text{c}\psi & \text{c}\phi\text{s}\theta\text{s}\psi - \text{c}\psi\text{s}\phi \\ -\text{s}\theta & \text{c}\theta\text{s}\phi & \text{c}\theta\text{c}\phi \end{pmatrix} \tag{3}$$

where θ, ϕ, and ψ are Euler angles representing pitch, roll, and yaw, respectively, and $\text{s}(\cdot), \text{c}(\cdot)$ denote the $\sin(\cdot)$ and $\cos(\cdot)$ functions. The relation between the angular velocity vector and the vector of Euler angles is given by

$$\begin{pmatrix} \dot{\psi} \\ \dot{\theta} \\ \dot{\phi} \end{pmatrix} = \mathscr{M}(\psi, \theta, \phi)\omega \tag{4}$$

where $\psi \in [-\pi, \pi), \theta \in (-\pi/2, \pi/2), \phi \in (-\pi/2, \pi/2)$, and

$$\mathscr{M}(\psi, \theta, \phi) = \begin{pmatrix} 0 & \text{s}\phi \sec\theta & \text{c}\phi \sec\theta \\ 0 & \text{c}\phi & -\text{s}\phi \\ 1 & \text{s}\phi \tan\theta & \text{c}\phi \tan\theta \end{pmatrix} \tag{5}$$

is invertible in the region where (ψ, θ, ϕ) are allowed to take values.

The propeller dynamics is reduced to linear relationships between the squared rotor angular rate and the force and moment of the propellers which are given by

$$F_{zi} = -b\varpi_i^2, \tag{6}$$

$$\tau_{zi} = \kappa\varpi_i^2, \tag{7}$$

respectively, where i is the motor index, b and κ are thrust and drag coefficients, and ϖ_i is the rotor angular velocity. All the motors are aligned with the vertical axis of the body frame. Indeed, force and moments occur in the z-direction of the body frame, where F_z is negative to be consistent with the downward z-axis convention. The mapping that links the motor torque inputs to the rigid body forces and moments is given by

$$\begin{pmatrix} F \\ \tau_x \\ \tau_y \\ \tau_z \end{pmatrix} = \begin{pmatrix} -b & -b & -b & -b \\ 0 & bl & 0 & -bl \\ bl & 0 & -bl & 0 \\ \kappa & -\kappa & \kappa & -\kappa \end{pmatrix} \begin{pmatrix} \varpi_1^2 \\ \varpi_2^2 \\ \varpi_3^2 \\ \varpi_4^2 \end{pmatrix} \tag{8}$$

where l represents the distance between the rotors with respect to the center of mass (center of gravity) of the quadrotor. The rotational velocity of the quadrotor, namely, ω and the sum of angular velocity of the rotors, namely, $\varpi_r = -\varpi_1 + \varpi_2 - \varpi_3 + \varpi_4$ determine the resultant gyroscopic moment, that is, $M_g = \omega \times J_r \varpi_r$. J_r represents the inertia of the rotating rotors, which is considered the same parameter for all four motors.

Considering that Euler angles are needed to attitude control, and also that from the translational coordinates only that for altitude is required, asumming low speeds, the equations of motion for a quadrotor are then given by Ref. [16]

$$\ddot{z} = -g + (c\theta c\phi)\frac{F}{m_q}, \tag{9}$$

$$\ddot{\theta} = \left(\frac{J_z - J_x}{J_y}\right)\dot{\phi}\dot{\psi} + \frac{J_r}{J_y}\varpi_r\dot{\phi} + \frac{l}{J_y}\tau_y, \tag{10}$$

$$\ddot{\phi} = \left(\frac{J_y - J_z}{J_x}\right)\dot{\theta}\dot{\psi} - \frac{J_r}{J_x}\varpi_r\dot{\theta} + \frac{l}{J_x}\tau_x, \tag{11}$$

$$\ddot{\psi} = \left(\frac{J_x - J_y}{J_z}\right)\dot{\theta}\dot{\phi} + \frac{1}{J_z}\tau_z, \tag{12}$$

where z is for the vertical dynamics. J_x, J_y, and J_z are moments of inertia in the direction of the three-dimensional Cartesian coordinates.

3. Quadrotor's decentralized parametric model

In this work, the proposal is centered in the design of a decentralized robust direct MRAC for the attitude of a quadrotor UAV. So, the mathematical model of the quadrotor (Eqs. (9)–(12)) should be expressed in a convenient form for its analysis in the searching of the main goal.

The first step toward that goal is the selection of a set of state variables as follows:

$$\begin{array}{llll} x_1 = \phi, & x_3 = \theta, & x_5 = \psi, & x_7 = z, \\ x_2 = \dot{x}_1 = \dot{\phi}, & x_4 = \dot{x}_3 = \dot{\theta}, & x_6 = \dot{x}_5 = \dot{\psi}, & x_8 = \dot{x}_7 = \dot{z}. \end{array} \tag{13}$$

Thus, the description of the model for the quadrotor (Eqs. (9)-(12)) in state-variable equations takes the form

$$\dot{x} = \begin{pmatrix} x_2 \\ \left(\frac{J_y - J_z}{J_x}\right) x_4 x_6 - \frac{J_r}{J_x} \varpi_r x_4 + \frac{l}{J_x} \tau_x \\ x_4 \\ \left(\frac{J_z - J_x}{J_y}\right) x_2 x_6 + \frac{J_r}{J_y} \varpi_r x_2 + \frac{l}{J_y} \tau_y \\ x_6 \\ \left(\frac{J_x - J_y}{J_z}\right) x_2 x_4 + \frac{1}{J_z} \tau_z \\ x_8 \\ -g + (cx_1 cx_3) \frac{F}{m_q} \end{pmatrix}. \tag{14}$$

Moreover, neglecting the gyroscopic torques due to Euler angles change slightly during flying, the rotational subsystem can be subdivided as follows:

$$\dot{x}^1 = \begin{pmatrix} x_2 \\ \left(\frac{J_y - J_z}{J_x}\right) x_4 x_6 + \frac{l}{J_x} \tau_x \end{pmatrix}, \tag{15}$$

$$\dot{x}^2 = \begin{pmatrix} x_4 \\ \left(\frac{J_z - J_x}{J_y}\right) x_2 x_6 + \frac{l}{J_y} \tau_y \end{pmatrix}, \tag{16}$$

$$\dot{x}^3 = \begin{pmatrix} x_6 \\ \left(\frac{J_x - J_y}{J_z}\right) x_2 x_4 + \frac{1}{J_z} \tau_z \end{pmatrix}, \tag{17}$$

with $\dot{x}^1 = (\dot{x}_1 \quad \dot{x}_2)^T, \dot{x}^2 = (\dot{x}_3 \quad \dot{x}_4)^T$, and $\dot{x}^3 = (\dot{x}_5 \quad \dot{x}_6)^T$.

In order to achieve the goal of attitude control task for the quadrotor UAV *via* MRAC design approach in a decentralized way, Eq. (15) is parameterized in the form

$$\dot{x}^1 = A_1 x^1 + B_1 \Lambda_1 (u^1 + \Theta_1 \Phi_1(x)) + \zeta_1 \tag{18}$$

with

$$A_1 = \begin{bmatrix} 0 & 1 \\ 0 & 0 \end{bmatrix}, \quad B_1 = \begin{bmatrix} 0 \\ 1 \end{bmatrix}, \tag{19}$$

$x^1 = (x_1 \quad x_2)^T, a_1 = \left(J_y - J_z\right)/J_x, b_1 = l/J_x, \Lambda_1 = b_1, \Theta_1 = a_1/b_1, \Phi_1(x) = x_4 x_6$, and $u^1 = \tau_x$. In a similar way, the parameterization for Eq. (16) is also given in the form

$$\dot{x}^2 = A_2 x^2 + B_2 \Lambda_2 (u^2 + \Theta_2 \Phi_2(x)) + \zeta_2 \tag{20}$$

where $A_2 = A_1, B_2 = B_1, x^2 = (x_3 \quad x_4)^T, a_2 = (J_z - J_x)/J_y, b_2 = l/J_y, \Lambda_2 = b_2$, $\Theta_2 = a_2/b_2, \ \Phi_2(x) = x_2 x_6$, and $u^2 = \tau_y$. Thus, for Eq. (17)

$$\dot{x}^3 = A_3x^3 + B_3\Lambda_3\big(u^3 + \Theta_3\Phi_3(x)\big) + \zeta_3 \tag{21}$$

with $A_3 = A_1, B_3 = B_1, x^3 = (x_5 \quad x_6)^T, a_3 = \left(J_x - J_y\right)/J_z, b_3 = 1/J_z, \Lambda_3 = b_3, \Theta_3 = a_3/b_3,\ \Phi_3(x) = x_2x_4$, and $u^3 = \tau_z$.

In the state space representation, the reference roll (model) dynamics is given by

$$\begin{pmatrix} \dot{x}_{1m} \\ \dot{x}_{2m} \end{pmatrix} = \begin{pmatrix} 0 & 1 \\ -\omega_n^2 & -2\xi\omega_n \end{pmatrix}\begin{pmatrix} x_{1m} \\ x_{2m} \end{pmatrix} + \begin{pmatrix} 0 \\ \omega_n^2 \end{pmatrix} x_{1r}, \tag{22}$$

which comes from the transfer function formulation for a second-order system model, with x_{1m} the reference angle, x_{1r} the command signal, ω_n the desired natural frequency, and ξ the damping ratio.

In general, given $(A, B, \Lambda, A_m, B_m)$, there is no guarantee that ideal gains K_x and K_r satisfy the matching conditions

$$A + B\Lambda K_x^T = A_m, \tag{23}$$

$$B\Lambda K_r^T = B_m. \tag{24}$$

From the experience, the structure of A is known so the pair (A_m, B_m) may be chosen such that (Eqs. (23) and (24)) have at least one solution pair (K_x, K_r).

3.1 Direct MRAC under unmatched external perturbations

It is desirable that for every subsystem (Eqs. (18)-(21)) the closed-loop dynamics have the form

$$\dot{x} = A_m x + B\Lambda\big(u + \Theta^T\Phi(x)\big) + B_m r + \zeta(t). \tag{25}$$

So, choosing the control input

$$u = -\hat{\Theta}^T\Phi(x) \tag{26}$$

and substituting Eq. (26) in Eq. (25) it yields

$$\dot{x} = A_m x - B\Lambda\tilde{\Theta}^T\Phi(x) + B_m r + \zeta(t) \tag{27}$$

with

$$\tilde{\Theta}^T = \hat{\Theta}^T - \Theta^T \tag{28}$$

the parameter estimation error.

Let us consider the tracking error equation

$$e = x - x_m. \tag{29}$$

Subtracting the reference model dynamics, that is,

$$\dot{x}_m = A_m x_m + B_m r, \tag{30}$$

from Eq. (25), it yields the error dynamics

$$\dot{e} = A_m e - B\Lambda\tilde{\Theta}^T\Phi(x) + \zeta(t). \tag{31}$$

In order to guarantee uniform ultimate boundedness (UUB) of all signals in the closed-loop dynamics, considering the Lyapunov function candidate

$$V(e, \tilde{\Theta}) = e^T P e + \mathrm{tr}\left(\tilde{\Theta}^T \Gamma_\Theta^{-1} \tilde{\Theta}\Lambda\right), \tag{32}$$

where $\Gamma_\Theta = \Gamma_\Theta^T > 0$ denotes the adaptation rate and $P = P^T > 0$ is the unique symmetric positive-definite solution of the Lyapunov equation, computing its time derivative along the trajectories of the tracking error dynamics (Eq. (31)) yields

$$\dot{V}(e, \tilde{\Theta}) = -e^T Q e - 2e^T PB\Lambda\tilde{\Theta}^T\Phi(x) + 2e^T P\zeta(x) + 2\mathrm{tr}\left(\tilde{\Theta}^T \Gamma_\Theta^{-1}\dot{\hat{\Theta}}\Lambda\right), \tag{33}$$

which can be rewritten as

$$\dot{V}(e, \tilde{\Theta}) = -e^T Q e + 2\mathrm{tr}\left(\tilde{\Theta}^T\left\{\Gamma_\Theta^{-1}\dot{\hat{\Theta}} - \Phi e^T PB\right\}\Lambda\right) + 2e^T P\zeta(t). \tag{34}$$

Thus, considering the update law

$$\dot{\hat{\Theta}} = \Gamma_\Theta \Phi(x) e^T PB \tag{35}$$

the time derivative $\dot{V}$ of V then results

$$\dot{V}(e, \tilde{\Theta}) = -e^T Q e + 2e^T P\zeta(t) \leq -\lambda_{\min}(Q)\|e\|^2 + 2\|e\|\lambda_{\max}(P)\zeta_{\max} \tag{36}$$

and, consequently, $\dot{V} < 0$ outside of the set

$$E_0 = \left\{(e, \tilde{\Theta}) : \|e\| \geq 2\frac{\lambda_{\max}(P)}{\lambda_{\min}(Q)}\zeta_{\max} = e_0\right\}. \tag{37}$$

The trajectories $e(t)$ of the error dynamics (Eq. (31)) enter a compact set $(\Omega_0 \supset E_0) \subset R^n$ in finite time and will remain there for all future times. However, Ω_0 is not compact in the $(e, \tilde{\Theta})$ space. Indeed, Ω_0 is unbounded since the parameter estimation error $\tilde{\Theta}$ are not restricted at all. Thus, inside Ω_0, $\dot{V}$ can become positive and the parameter error $\tilde{\Theta}$ may diverge, even though the tracking error norm remains finite at all times. This phenomenon is known as *parameter drift*, and it is caused by the perturbation term $\zeta(t)$. This argument claims that the updated law (Eq. (35)) is not robust to bounded perturbations.

3.2 Robust direct MRAC under unmatched external perturbations

From Eq. (35), in order to gain robustness the updated law

$$\dot{\hat{\Theta}} = \Gamma_\Theta\left(\Phi(x)e^T PB - \sigma\hat{\Theta}\right) \tag{38}$$

is chosen, with an extra term in order to add damping and σ a strictly positive constant.

Substituting Eq. (38) in Eq. (34), after some operations results

$$\dot{V}\left(e,\tilde{\Theta}\right) = -e^T Q e - 2\sigma \mathrm{tr}\left(\tilde{\Theta}^T \hat{\Theta} \Lambda\right) + 2e^T P \zeta(t). \tag{39}$$

From Eq. (28), Eq.(39) can be written as

$$\dot{V}\left(e,\tilde{\Theta}\right) = -e^T Q e - 2\sigma \mathrm{tr}\left(\tilde{\Theta}^T \tilde{\Theta} \Lambda\right) - 2\sigma \mathrm{tr}\left(\tilde{\Theta}^T \Theta \Lambda\right) + 2e^T P \zeta(t). \tag{40}$$

By definition,

$$\mathrm{tr}\left(\tilde{\Theta}^T \tilde{\Theta} \Lambda\right) = \sum_{i=1}^{N} \sum_{j=1}^{m} \tilde{\Theta}_{ij}^2 \Lambda_{ii} \geq \|\tilde{\Theta}\|_F^2 \Lambda_{\min} \tag{41}$$

with $\|\tilde{\Theta}\|_F^2 = \sum_{i=1}^{N} \sum_{j=1}^{m} \tilde{\Theta}_{ij}^2$ the Frobenius norm and $\Lambda_{\min}$ the minimum diagonal element of Λ. Also, from the Schwarz inequality,

$$|\mathrm{tr}\left(\tilde{\Theta}^T \Theta \Lambda\right)| \leq \|\tilde{\Theta}^T \Theta\|_F \|\Lambda\|_F \leq \|\tilde{\Theta}\|_F \|\Theta\|_F \|\Lambda\|_F \tag{42}$$

Substituting Eqs. (41) and (42) into Eq. (40) yields

$$\begin{aligned}\dot{V}\left(e,\tilde{\Theta}\right) \leq & -\lambda_{\min}(Q)\|e\|^2 + 2\|e\|\lambda_{\max}(P)\zeta_{\max} - 2\sigma\|\tilde{\Theta}\|_F^2 \Lambda_{\min} \\ & +2\sigma\|\tilde{\Theta}\|_F \|\Theta\|_F \|\Lambda\|_F.\end{aligned} \tag{43}$$

The first and second terms, by completion of squares, can be rewritten as

$$-\lambda_{min}(Q)\|e\|^2 + 2\|e\|\lambda_{\max}(P)\zeta_{\max} = -\lambda_{\min}(Q)\left(\|e\| - \frac{\lambda_{\max}(P)}{\lambda_{\min}(Q)}\zeta_{\max}\right)^2 + \frac{\lambda_{\max}^2(P)}{\lambda_{\min}(Q)}\zeta_{\max}^2 \tag{44}$$

In a similar way, the remaining terms can be rewritten as

$$-2\sigma\|\tilde{\Theta}\|_F^2 \Lambda_{\min} + 2\sigma\|\tilde{\Theta}\|_F \|\Theta\|_F \|\Lambda\|_F = -2\sigma\Lambda_{\min}\left(\|\tilde{\Theta}\|_F - \frac{1}{2}\|\Theta\|_F \frac{\|\Lambda\|_F}{\Lambda_{\min}}\right)^2 + \sigma\frac{\|\Theta\|_F^2 \|\Lambda\|_F^2}{2\Lambda_{\min}} \tag{45}$$

Then, (Eqs. (43)) can be written as

$$\dot{V}\left(e,\tilde{\Theta}\right) \leq -\lambda_{\min}(Q)\left(\|e\| - \frac{\lambda_{\max}(P)}{\lambda_{\min}(Q)}\zeta_{\max}\right)^2 + \frac{\lambda_{\max}^2(P)}{\lambda_{\min}(Q)}\zeta_{\max}^2 \tag{46}$$

$$-2\sigma\Lambda_{\min}\left(\|\tilde{\Theta}\|_F - \frac{1}{2}\|\Theta\|_F \frac{\|\Lambda\|_F}{\Lambda_{\min}}\right)^2 + \sigma\frac{\|\Theta\|_F^2 \|\Lambda\|_F^2}{2\Lambda_{\min}} \tag{47}$$

Consequently, $\dot{V}\left(e,\tilde{\Theta}\right) < 0$ if

$$\|e\| > \sqrt{\frac{1}{\lambda_{\min}(Q)}\left(\frac{\lambda_{\max}^2(P)}{\lambda_{\min}(Q)}\zeta_{\max}^2 + \sigma\frac{\|\Theta\|_F^2\|\Lambda\|_F^2}{2\Lambda_{\min}}\right)} + \frac{\lambda_{\max}(P)}{\lambda_{\min}(Q)}\zeta_{\max} = c_1 \tag{48}$$

or

$$\|\tilde{\Theta}\|_F > \sqrt{\frac{1}{2\sigma\Lambda_{\min}}\left(\frac{\lambda_{\max}^2(P)}{\lambda_{\min}(Q)}\zeta_{\max}^2 + \sigma\frac{\|\Theta\|_F^2\|\Lambda\|_F^2}{2\Lambda_{\min}}\right)} + \frac{1}{2}\|\Theta\|_F\frac{\|\Lambda\|_F}{\Lambda_{\min}} = c_2 \tag{49}$$

Strictly speaking, $\dot{V}(e,\tilde{\Theta}) < 0$ outside of the compact set, closed and bounded,

$$\Omega = \left\{(e,\tilde{\Theta}) : (\|e\| \le c_1)\wedge(\|\tilde{\Theta}\|_F \le c_2)\right\} \subset R^n \times R^{N\times m}. \tag{50}$$

So, Eq. (50) proves UUB tracking of the external command r by the system output $y(t)$. Thus, command tracking is achieved in the presence of parametric uncertainties (Λ, Θ) and nonparametric bounded time-varying perturbations $\zeta(t)$. Then, the command tracking problem for the MIMO system dynamics (Eq. (25)) is solved, and it is formulated as follows.

Theorem 1. Given MIMO dynamics (Eq. (18)) with a control uncertainty Λ_1 and a matched unknown function $f_1(x) = \Theta_1^T\Phi_1(x)$, the MRAC subsystem (Eqs. (26), (28), (29), (38)) enforces uniform ultimate bounded tracking performance of the reference model dynamics (Eq.(30)), driven by any bounded time-varying command $r(t)$. Moreover, all signals in the corresponding closed-loop subsystem remain UUB in time.

4. Simulation results

In what follows, simulation results about the performance of the decentralized robust direct MRAC for the attitude of a quadrotor UAV are discussed. The proposal is evaluated when considering both the absence and the presence of perturbations. The perturbations come from a Gaussian noise source. In this work, the trajectory to be followed is demanding and the simulation time is long in order to evaluate the drift of the parametric estimates. From Refs. [17, 18], with $J_x = J_y = 0.03\,\text{kg m}^2$, $J_z = 0.04\,\text{kg m}^2$, $l = 0.2\,\text{m}$, $m = 1.79\,\text{kg}$, and selecting $\omega_n^2 = 32.653\,\text{rad/s}$ and $\xi = 0.7$, the matching conditions (Eqs. (23) and (24)) are then satisfied to ideal gains $K_{x^1} = (-4.89795 \quad -1.2)^T$ and $K_{r^1} = 4.89795$.

By selecting

$$Q = \begin{pmatrix} 1 & 0 \\ 0 & 10 \end{pmatrix} \tag{51}$$

and solving the algebraic Lyapunov equation

$$PA_m + A_m^T P = -Q \tag{52}$$

it yields

$$P = \begin{pmatrix} 20.5931 & 0.0153 \\ 0.0153 & 0.6269 \end{pmatrix}. \tag{53}$$

After several iterations, the adaptation rate for the update law (Eq. (38)) is given by $\Gamma_{\Theta_1} = 9500$. Since the same MRAC design approach is also carried out in the achievement of the stabilization for pitch and yaw angles, only the design procedure for roll angle stabilization is shown. So, the same P matrix is obtained when considering the same values for ω_n, ξ, and Q matrix. Nevertheless, the remaining adaptation rates are given by $\Gamma_{\Theta_2} = 9430$ and $\Gamma_{\Theta_3} = 9500$.

It must be noticed that a PID controller is implemented for altitude control [19, 20].

From **Figures 1–5**, numerical simulation results from evaluating the viability of the proposal to the case where the quadrotor UAV is performing the model-following task to the attitude in an environment free of external perturbations are shown. From **Figures 6–13**, simulation results are shown but to the case for which the model-following task to the quadrotor's attitude is carried out when considering the presence of perturbations.

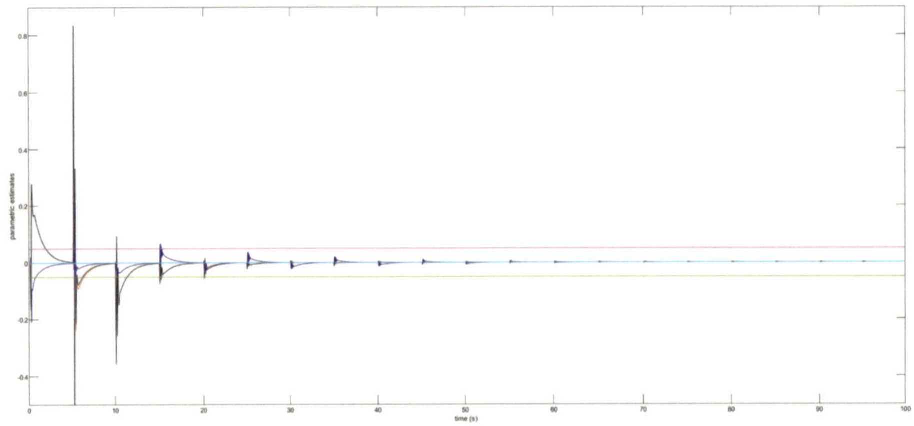

Figure 1.
Parameter estimates $\hat{\Theta}_1, \hat{\Theta}_2$ and $\hat{\Theta}_3$ in the absence of perturbations.

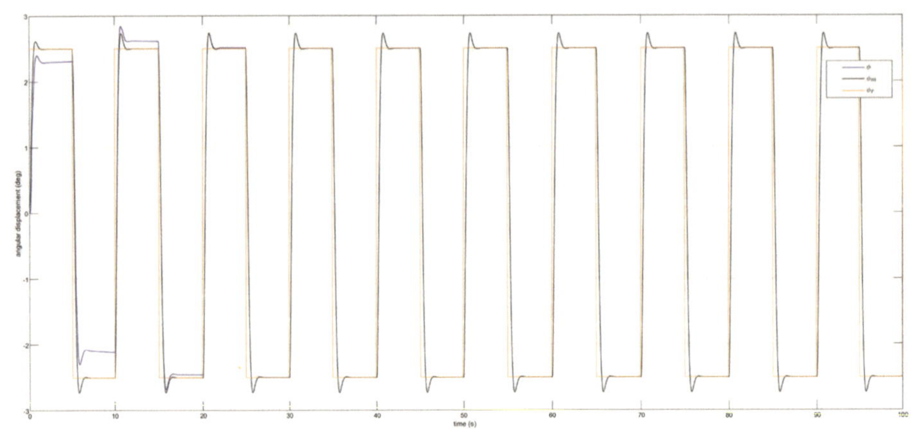

Figure 2.
Model following by the roll angle in the absence of perturbations.

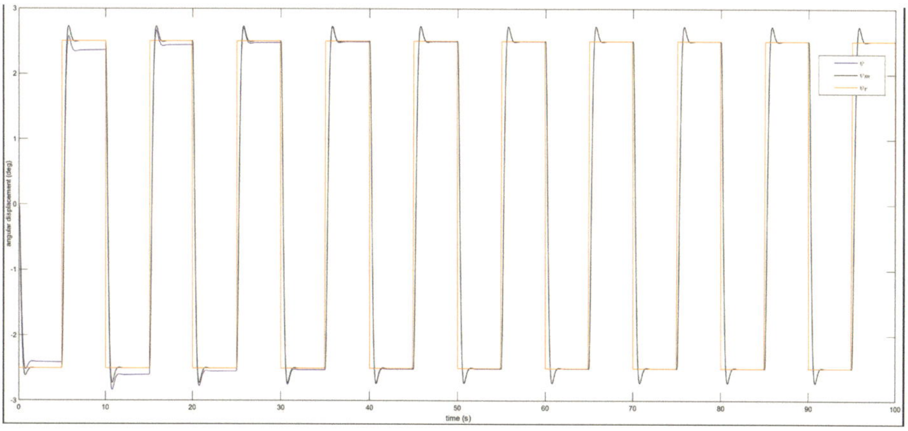

Figure 3.
Model following by the pitch angle in the absence of perturbations.

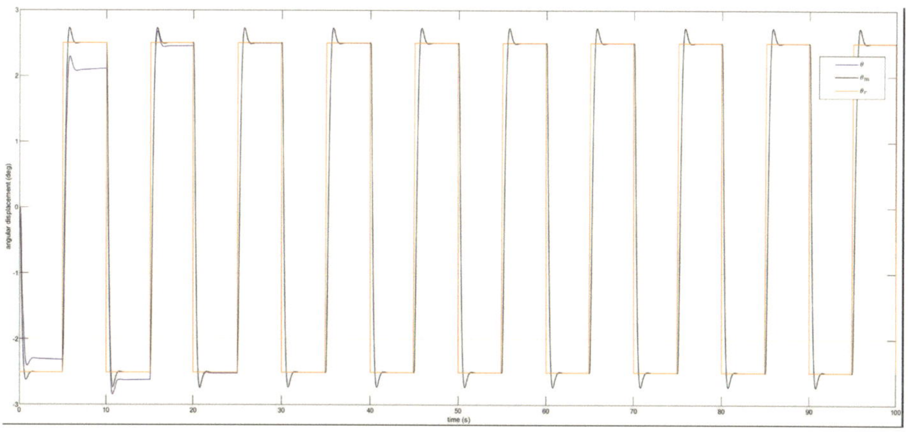

Figure 4.
Dynamics of the yaw angle in the absence of perturbations.

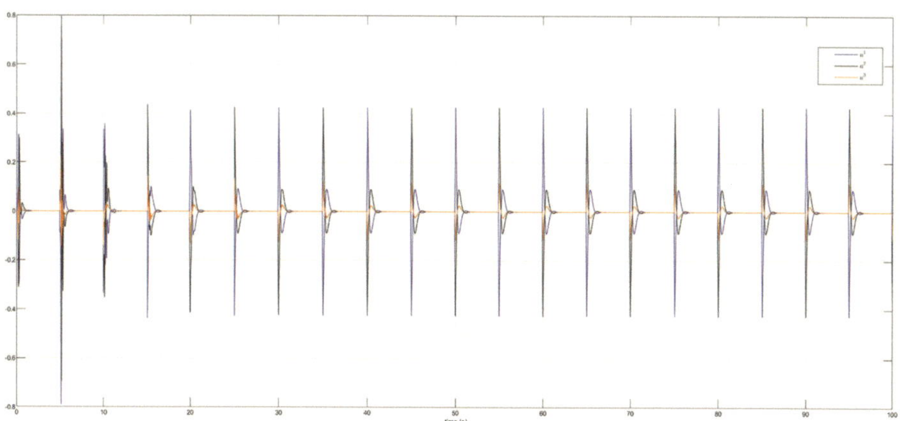

Figure 5.
Dynamics from the adaptive control signals u^1, u^2 and u^3 in the absence of perturbations.

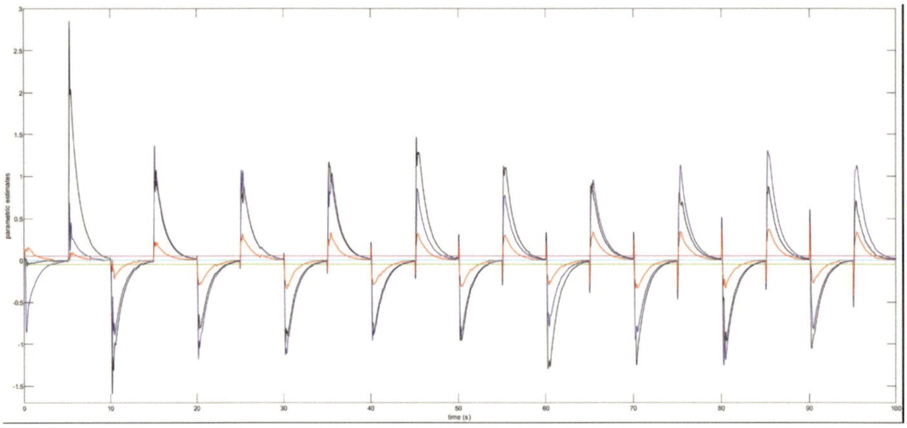

Figure 6.
Parameter estimates $\hat{\Theta}_1, \hat{\Theta}_2$ and $\hat{\Theta}_3$ in the presence of perturbations.

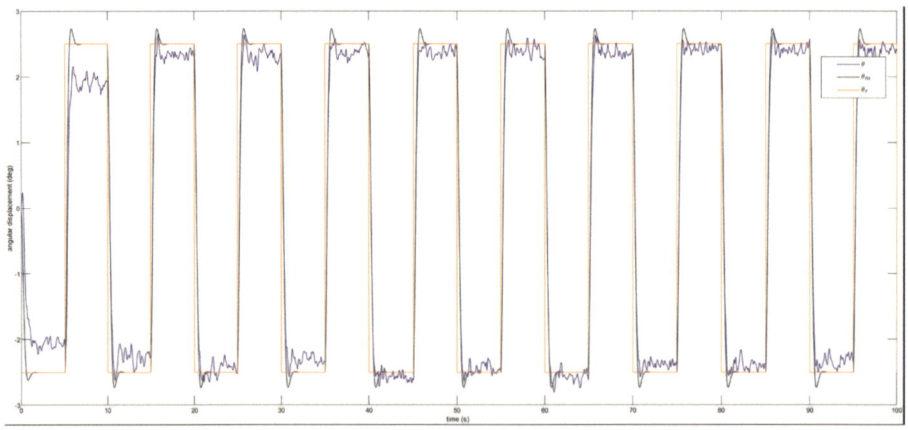

Figure 7.
Model following of the roll angle under the presence of perturbations.

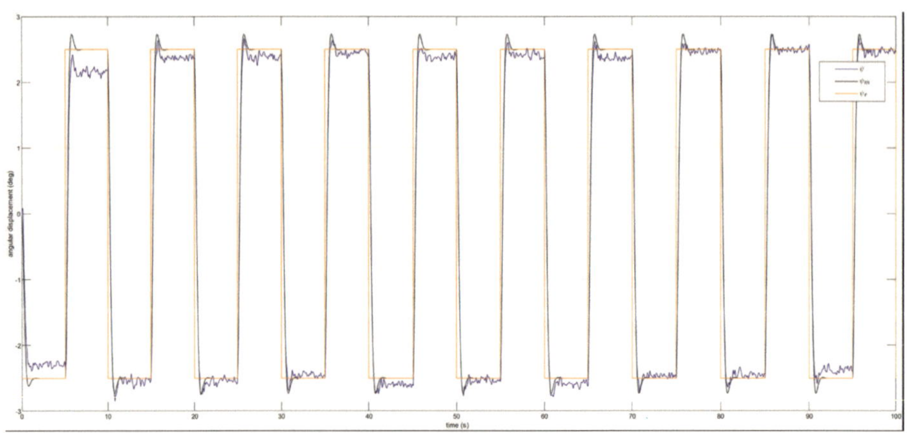

Figure 8.
Dynamics of the pitch angle in the presence of perturbations.

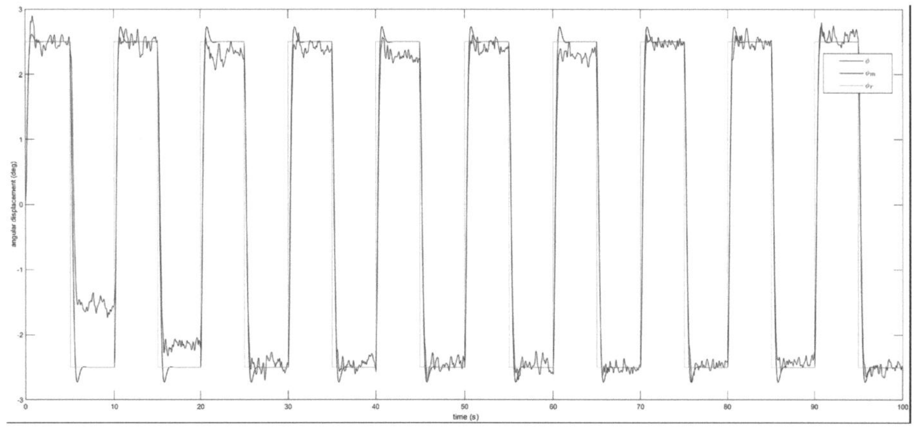

Figure 9.
Dynamics of the yaw angle when facing the presence of perturbations.

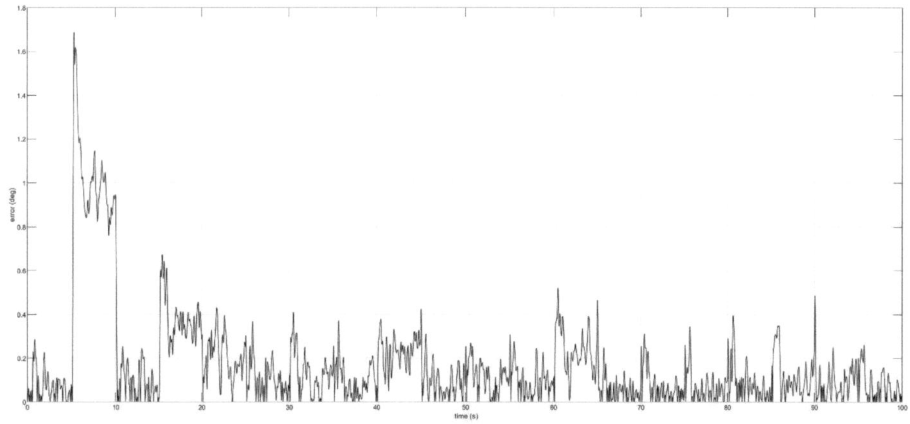

Figure 10.
Tracking error signal from the roll motion.

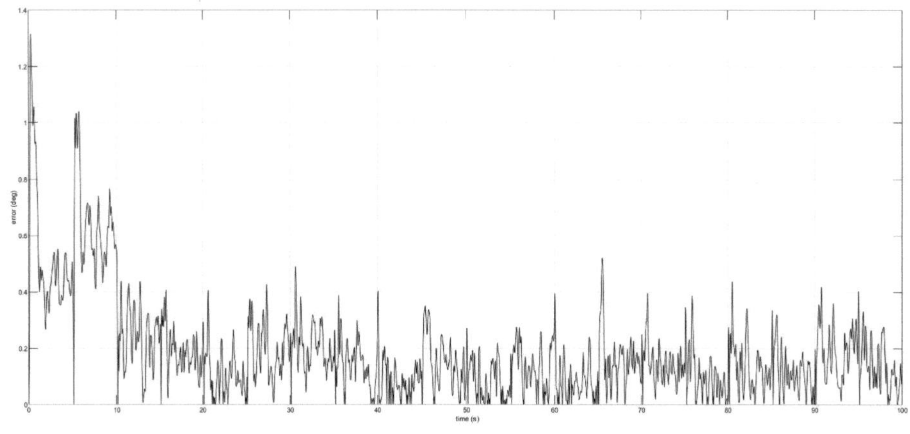

Figure 11.
Tracking error signal from the pitch motion.

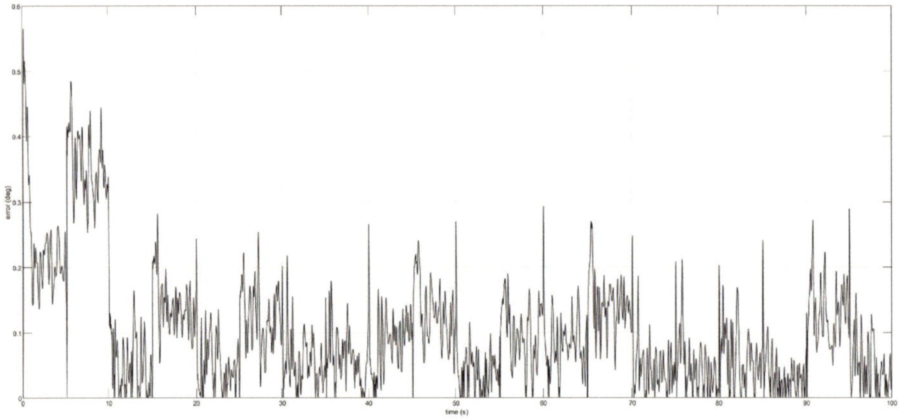

Figure 12.
Tracking error signal from the yaw angle.

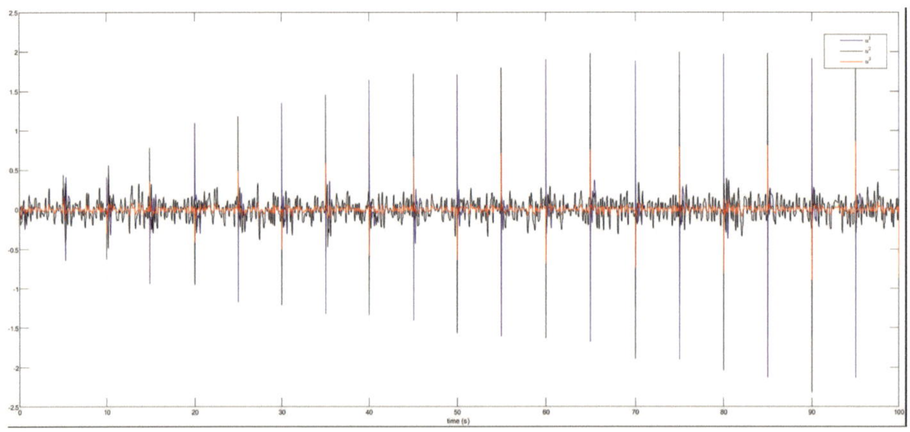

Figure 13.
Dynamics from the adaptive control signals u^1, u^2 and u^3 when confronting the perturbations.

The bounds and norms of the error signal from each decentralized subsystem are shown in **Table 1** for $\sigma = 0.0001$. So, from **Table 1**, it can be seen that the inequality (Eq. (48)) is satisfied. Thus, from Theorem 1, the command tracking problem for the MIMO system dynamics is solved. In contrast with other robust learning algorithms, where prior knowledge of an upper bound ζ_{max} for the system perturbation $\zeta(t)$ is assumed, in our proposal any prior information on the system perturbation upper bounds is not required.

Subsystem	c_1	$\|e\|$
Roll motion	10.29	26.02
Pitch motion	10.29	24.11
Yaw motion	10.29	12.65

Table 1.
Bounds and norms of the error signal from each decentralized subsystem.

5. Conclusions

From simulation results, it can be seen that the parameter estimates do not drift to infinity when the decentralized direct MRAC has to deal with the perturbations. So, it can be concluded that the decentralized direct MRAC for the attitude of the quadrotor UAV is robust. Also, it must be highlighted that in the absence of external perturbations, the parametric estimates converge to the origin. Moreover, it should be noticed that for both cases, the quadrotor UAV reaches the desired height. Besides, it should be noticed that when the tracking error becomes small, the update law dynamics (Eq. (38)) can be approximated by $-\Gamma_\Theta \sigma \hat{\Theta}$. So, for small tracking errors, the adaptive parameters have a tendency to return to the origin; that is, they disregard the gain values that caused the tracking error to become small. Also, even if the perturbation $\zeta(t)$ is removed from the system dynamics (Eq. (25)), and if the reference command r is persistently exciting, the parameter estimation errors (Eq. (28)) do not converge to the origin.

Acknowledgements

This research was funded by Tecnológico Nacional de México (TecNM) through project 20496.24-P and, partially, under grant 43433 from the EDD 2023 program.

Abbreviations

DC	direct current
MIMO	multiple-input multiple-output
MRAC	model reference adaptive control
PID	proportional-integral-derivative
UAV	unmanned aerial vehicle
UUB	uniform ultimate boundedness

Author details

Francisco Jurado
Tecnológico Nacional de México/Instituto Tecnológico de La Laguna, Torreón, México

*Address all correspondence to: fjuradoz@lalaguna.tecnm.mx

References

[1] Ioannou PA, Sun J. Robust Adaptive Control. Upper Saddle River NJ: Prentice Hall PTR; 1995. 821 p

[2] Lavretsky E, Wise KA. Robust and Adaptive Control with Aerospace Applications. London: Springer-Verlag; 2013. 454 p

[3] Ioannou P, Fidan B. Adaptive Control Tutorial. Philadelphia PA: SIAM; 2006. 389 p

[4] Peterson BB, Narendra KS. Bounded error adaptive control. IEEE Transactions on Automatic Control. 1982;**27**:1161-1168

[5] Ioannou PA, Kokotovic PV. Adaptive Systems with Reduced Models. New York: Springer; 1983. 172 p

[6] Åström KJ, Wittenmark B. Adaptive Control. 2nd ed. New York, NY, USA: Addison-Wesley; 1995. 574 p

[7] Narendra KS, Annaswamy AM. A new adaptive law for robust adaptive control without persistency of excitation. IEEE Transactions on Automatic Control. 1987;**32**:134-145

[8] Narendra KS, Annaswami AM. Stable Adaptive Systems. Englewood Cliffs NJ: Prentice Hall; 1989. 494 p

[9] Slotine JJE, Coetsee JA. Adaptive sliding controller synthesis for non-linear systems. International Journal of Control. 1986;**43**(6):1639-1651

[10] Farrell JA, Polycarpou MM. Adaptive Approximation Based Control. Hoboken NJ: Wiley; 2006. 420 p

[11] Bodson M. Adaptive Estimation and Control. Coppell TX: Independently published; 2020. 253 p

[12] Slotine JJE, Li W. Applied Nonlinear Control. Englewood Cliffs NJ: Prentice Hall; 1991. 461 p

[13] Nguyen NT. Model-Reference Adaptive Control a Primer. Switzerland: Springer International Publishing AG; 2018. 444 p

[14] Etkin B, Reid LLD. Dynamics of Flight, Stability and Control. 3rd ed. New York, NY, USA: John Wiley & Sons; 1996. 382 p

[15] Nijmeijer H, van der Schaft A. Nonlinear Dynamical Control Systems. New York: Springer-Verlag; 1990. 439 p

[16] Bouabdallah S, Murrieri P, Siegwart R. Design and control of an indoor micro quadrotor. In: Proc. of the IEEE Int. Conf. On Robotics & Automation. New Orleans, LA: IEEE; 2004

[17] User Manual QBall 2 for QUARC. Ontario, Canada: Quanser Inc.; 2014

[18] Jurado F, Lopez S. A wavelet neural control scheme for a quadrotor unmanned aerial vehicle. Philosophical Transactions of the Royal Society A. 2018;**376**:20170248

[19] Jurado F, Hernández R. Decentralized MRAC with integral action for attitude control of a quadrotor UAV. In: Proc. of the, 2018 IEEE Int. Autumn Meeting on Power, Electronics and Computing (ROPEC), Ixtapa, Mexico. New York, NY, USA: IEEE; 2018

[20] Åström K, Hagglund T. PID Controllers: Theory, Design, and Tuning. 2nd ed. ISA - The Instrumentation, Systems and Automation Society; 1995. 343 p

Chapter 6

Perspective Chapter: Intelligent Adaptive Flight Control Augmentation of a Dynamic Inversion Autopilot

Quang M. Lam, Anthony J. Calise and Nhan T. Nguyen

Abstract

High-speed UAV flight control is challenged by unknown external aerodynamic disturbances and internal system variations due to complex aerodynamic configuration, propellant consumption, the center of gravity movement, and possible actuator failures. Redundant aerodynamic control effectors and engine inlet control mechanisms are exploited/blended to maintain adequate force/moment control to satisfy flight control reliability requirements. A control allocation (CA) mixing matrix is an essential element of the redundancy design in addressing fault tolerance capabilities. This paper employs an Intelligent Flight Controller (IFC) implemented in an adaptive control augmentation fashion to assist the UAV's primary Nonlinear Dynamic Inversion (NDI) controller in restoring aircraft stability and command following objective when subjected to effector performance degradation, including failures. Through a blended design between optimal control modification (OCM) and derivative-free model reference adaptive control (DF-MRAC), the IFC offers performance consistency in comparison to traditional MRAC. The proposed IFC framework has demonstrated its effectiveness in assisting the baseline NDI flight control system to maintain its mission subjected to actuator's failures (via an implicit automatic CA 're-distributing' action). Furthermore, the IFC also works well with any existing onboard CA algorithm in dealing with effector failures without requiring restructuring of the CA blending matrix. It therefore deserves consideration for application to future high Mach UAVs.

Keywords: nonlinear dynamic inversion (NDI), control allocation (CA), optimal control modification (OCM), derivative-free model reference adaptive control (DF-MRAC), direct adaptive control (DAC), passive CA, fault tolerant control (FTC)

1. Introduction

Flight control systems (FCSs) are one of the most important safety-critical systems in modern aircraft and high-speed UAVs. Redundancy of both FCS software (FSW) and its associated actuators/effectors (aerodynamic control surfaces and engine inlet control mechanisms) are exploited to achieve robust performance and high reliability against component failures (*hardware and/or software redundancy*, e.g., see [1, 2]). For hardware redundancy, multiple control effectors are designed and sized to adequately produce 6 degree-of-freedom (DOF) force/moment (F/M) responses in order to closely follow the commanded F/M profiles generated by the control law (CLAW) so that the aircraft can maintain its maneuvering ability subject to one or more effector failures. For example, with loss of one elevon and one canard out of a total of seven aerodynamic control surfaces, the aircraft may still be able to maintain its 6DOF controllability.

This paper investigates the employment of the intelligent flight control (IFC) architecture developed in [3, 4] for a generic transport model (GTM) aircraft, and further adopts, modifies, and implements the IFC in a control augmentation fashion to assist the Aerodynamic Model in Research Environment (ADMIRE) fighter aircraft FCS [5] in effectively maintaining its 6DOF FCS performance when subjected to two stuck inner elevons. It is shown that with the baseline dual loop NDI-FCS alone, the aircraft fails to maintain its desired flight mission, and consequently, its flight profile is prematurely terminated. The investigation studies herein also uncover several key connections or findings between adaptive control (CLAW side of FSW) and adaptive control allocation. One key connection between CLAW and CA is that the adaptive signal generated by the IFC is effective on the CLAW side in improving command tracking while improving CA actions in assigning which effectors should be utilized to achieve the required NDI F/M commands. Sections 2 and 5 of this paper will present these interesting details, while Sections 3 and 4 describe the problem statement of the baseline NDI controller subject to two stuck elevons and present the IFC formulation, respectively.

2. ADMIRE baseline controller description

The ADMIRE fighter jet and its development of a generic flight controller are presented in detail in [6, 7]. However, for completeness, it is briefly described here so its baseline FCS interface with the IFC block added in an adaptive control configuration can be accurately described and presented. The ADMIRE's generic dual loop design flight controller is employed in this study wherein the outer loop (i.e., slow dynamic loop) is designed to be command following. This amounts to achieving the angle of attack (AoA) command following for the longitudinal channel and for the angle of sideslip (AoS) and stability axis roll rate command following for the lateral and directional channels. The pitch/roll stick and rudder pedal commands are appropriately shaped and scaled to result in the AoA, stability axis roll rate, and AoS commands. A block schematic of the generic flight controller is shown in **Figure 1**. The inner loop (or fast dynamic loop) is designed using the Nonlinear Dynamic Inversion (NDI) adopted from [6] and detailed in the Appendix Section of the book for ADMIRE integration. The controller structure has the outer loop regulation of the slow dynamics consisting of the AoA, AoS, and flight path angle (FPA) by commanding the inner loop angular rates.

ADMIRE has seven aerodynamic control surfaces (CSs): left and right canards, left and right inboard elevons, left and right outboard elevons, and one rudder (see **Figure 2**). The canards and elevons can be operated together or in symmetric or differential deflections or individually. The roll, pitch, and yaw channels are controlled using these control surfaces. Hence, the transformation matrix has dimension

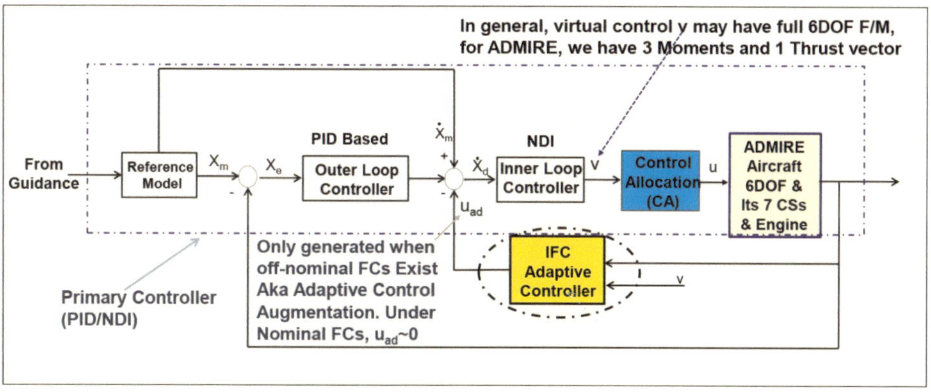

Figure 1.
ADMIRE dual loop autopilot with IFC implementation in an augmentation fashion – Baseline PID/NDI not using full state feedback.

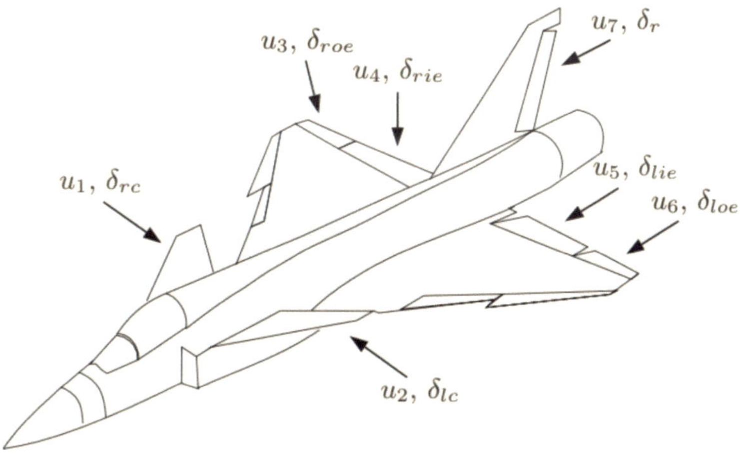

Figure 2.
ADMIRE's seven aerodynamic control surfaces (u_i, $i = 1,2,\dots 7$).

seven by three (three desired control moments transformed to seven CS deflections). The dual loop PID/NDI autopilot is detailed in [8] for readers who are interested in seeing how the commanded state vectors of both the outer loop and inner loops are being formulated. This paper is intended to focus on the IFC algorithm description and how it differs from [4] in the implementation. Note that the current ADMIRE simulation has a single engine; however, the engine throttle controller has not been jointly integrated with the seven CSs main PID/NDI controller. Nonetheless, it is being accounted for in the existing seven CA algorithms block (see **Figure 3** and [6] for algorithm details) to jointly map 'virtual' Force/Moment (F/M) command to 'real' effectors. For future UAV missions, especially for high-performance UAVs with flight speeds beyond Mach 5, a high-dimension effectors vector including advanced engine control effectors can be defined to address a more effective mixing scheme.

For the purposes of the roll axis control (the first column of B_{v2r} matrix, *virtual to real (v2r)*), the differential canards (KR2), and differential elevons (KR4) are used. For the pitch axis control (second column of B_{v2r} matrix, the symmetric canards (KP1) and symmetric elevons (KP3) are used (see [6, 7] for background). Finally, for the

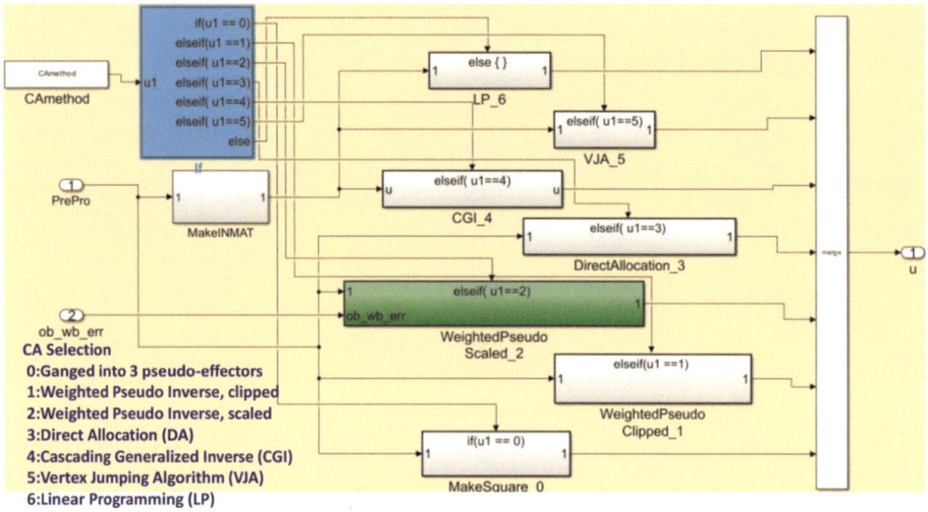

Figure 3.
Current ADMIRE CA block with existing seven CA algorithms and IFC works quite well with any of those seven without imposing any restructuring CA action.

yaw axis control (third column of B matrix), the differential canards (KY2), differential elevons (KY4), and rudder (KY5) are used. The control allocation matrix is given by:

$$B_{v2r} = \begin{bmatrix} -KR_2 & KP_1 & -KY_2 \\ KR_2 & KP_1 & KY_2 \\ -KR_4 & KP_3 & -KY_4 \\ -KR_4 & KP_3 & -KY_4 \\ KR_4 & KP_3 & KY_4 \\ KR_4 & KP_3 & KY_4 \\ KR_5 & KP_5 & KY_5 \end{bmatrix} \tag{1}$$

- KR5 is the aileron to rudder interconnect gain. In case the optimizer finds a solution greater in magnitude than 1.2, it is reset to 1.2. The maximum differential deflection of the elevons is 25 degrees, whereas that of the rudders is 30 degrees. Therefore, limiting the KR5 gain to 1.2 prevents a control surface saturation of the rudder due to a roll command at the expense of some sideslip buildup during roll maneuver about the stability axis.

- The maximum roll rate in the stability axis (velocity vector direction) is computed using the control allocation matrix determined above and used for the forward path (body x-axis direction) command scaling throughout the airspeed range.

3. Impact of two stuck inboard Elevons on the baseline NDI-FCS

In addition to the phase and gain margins, the following should be taken into consideration to ensure HQ criteria are met: (1) Effector performance degradation due to wear/tear or unexpected damage due to adversarial actions; (2) effector deflection and rate limits which could destabilize UAV FCS performance (see [9, 10]); (3) the AoA rate $\dot{\alpha}$ which should be safeguarded and not exceed an upper bound (e.g., < 25 deg./s) throughout the flight envelope (with an AoA bound not to exceed 40 deg); (4) full stick maximum roll acceleration 'Limit Bounds' violation; (5) AoS limit during roll maneuvers; (6) model following errors bound violation (e.g., see [10]). One of the key concepts going forward to satisfy high-speed UAV mission requirements is to jointly optimize F/M at the full 6DOF CLAW level as full 6DOF (currently ADMIRE has not met this design goal) to compute the desired F/M demand (whether a completely healthy set of effectors or a reduced set of degraded capacity effectors exists) and to produce a realistic, attainable F/M response.

Figures 4 and **5** present the ADMIRE Baseline FCS CS and trajectory performance under nominal operating conditions with seven healthy CSs (RC: Right Canard, LC: Left Canard, ROE: Right Outboard Elevon, RIE: Right Inboard Elevon, LIE: Left Inboard Elevon, LOE: Left Outboard Elevon, and RUD: Rudder).

The ADMIRE FCS performance when two inner elevons are stuck at and beyond 200 seconds is shown in **Figures 6** and 7. Under this failure, the ADMIRE flight trajectory is prematurely terminated at 433 seconds, as shown in **Figure 8**. The angular acceleration, angular rate, and attitude commands following comparisons are presented in **Figures 9–11**.

These figures demonstrate that with two stuck IB elevons, the ADMIRE fighter aircraft FCS is not able to complete its mission. As a result, modern adaptive control augmentation of some fashion is needed to augment/assist the primary baseline FCS to

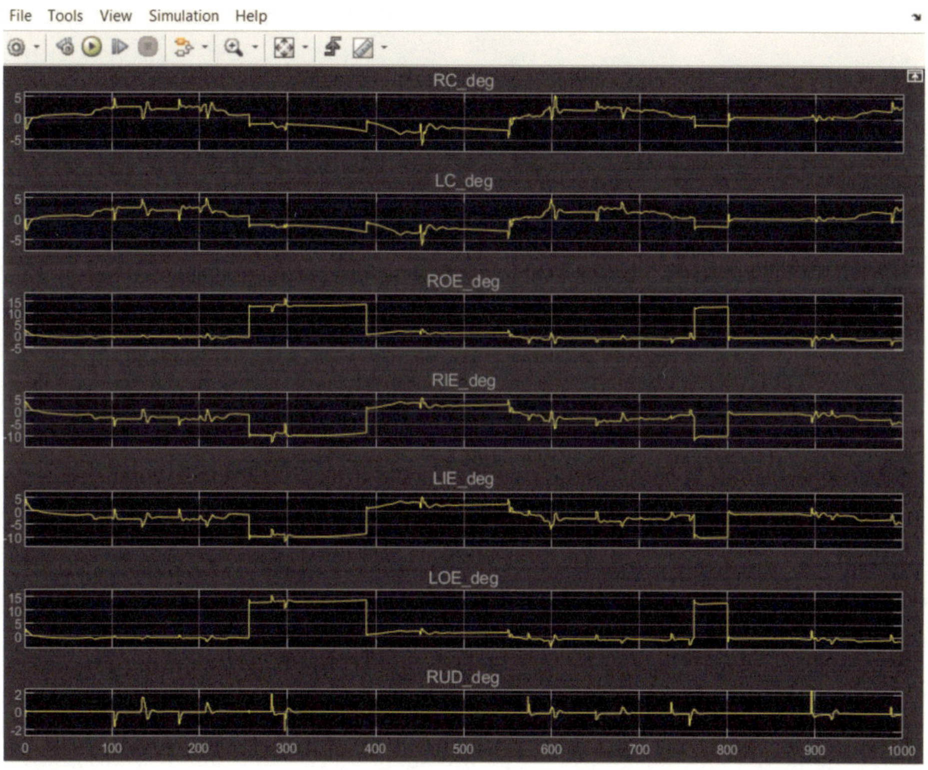

Figure 4.
7 CSs deflections under nominal FCs.

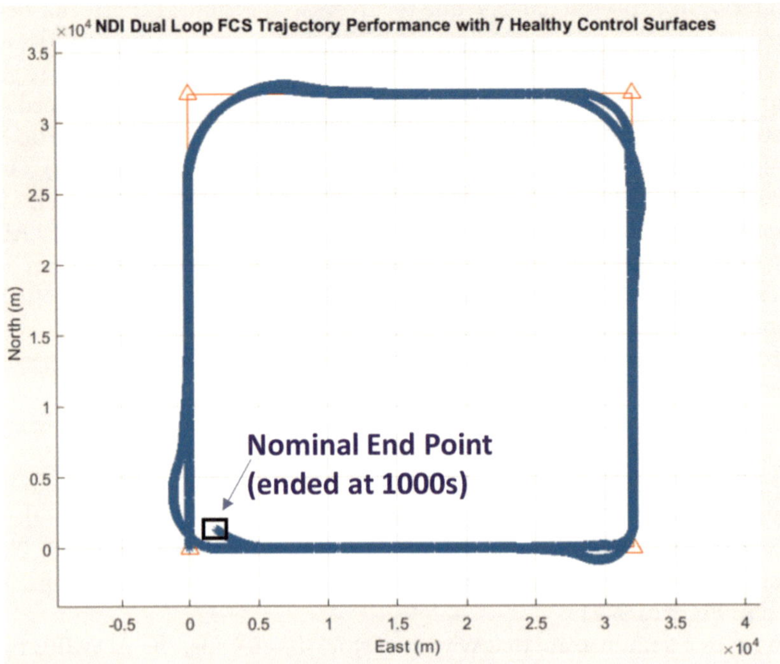

Figure 5.
ADMIRE nominal flight trajectory performance with seven healthy control surfaces.

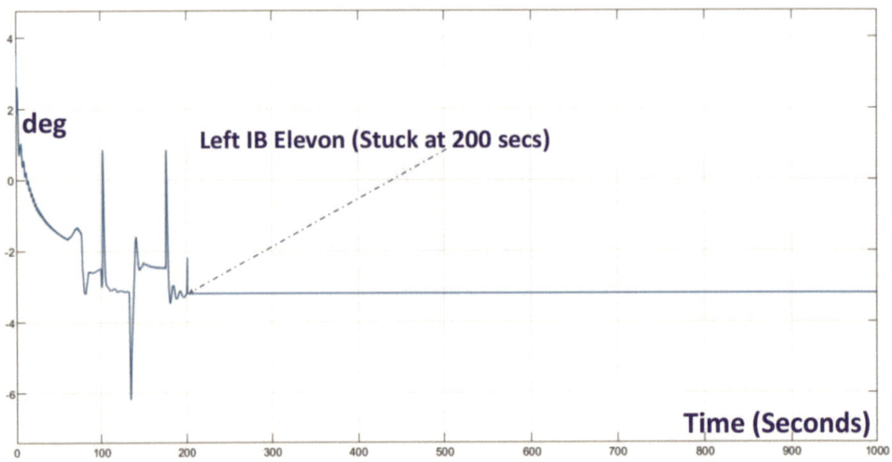

Figure 6.
Left inboard (IB) Elevon (stuck at 200 seconds).

cope with degraded flight conditions. In Section 4, we describe such an adaptive control augmentation concept using a hybrid adaptive control framework to best assist the primary FCS in completing its missions when subjected to unknown uncertainties and control effector failures.

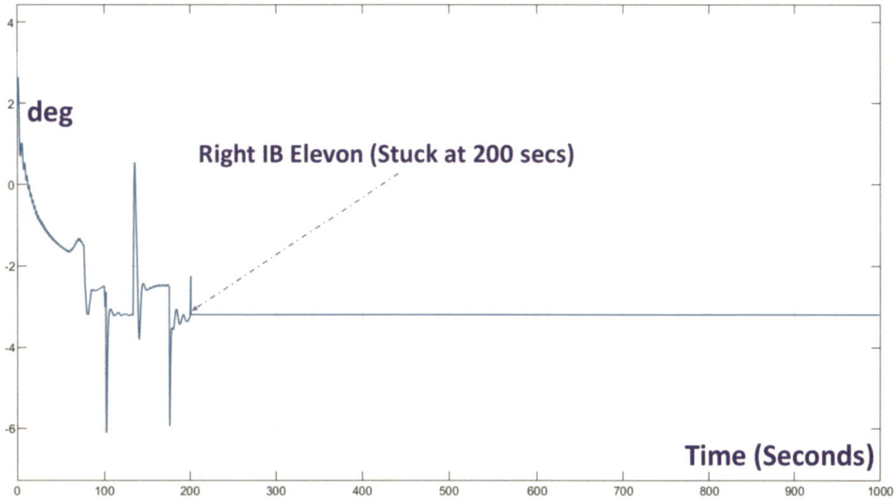

Figure 7.
Right inboard (IB) Elevon (stuck at 200 seconds).

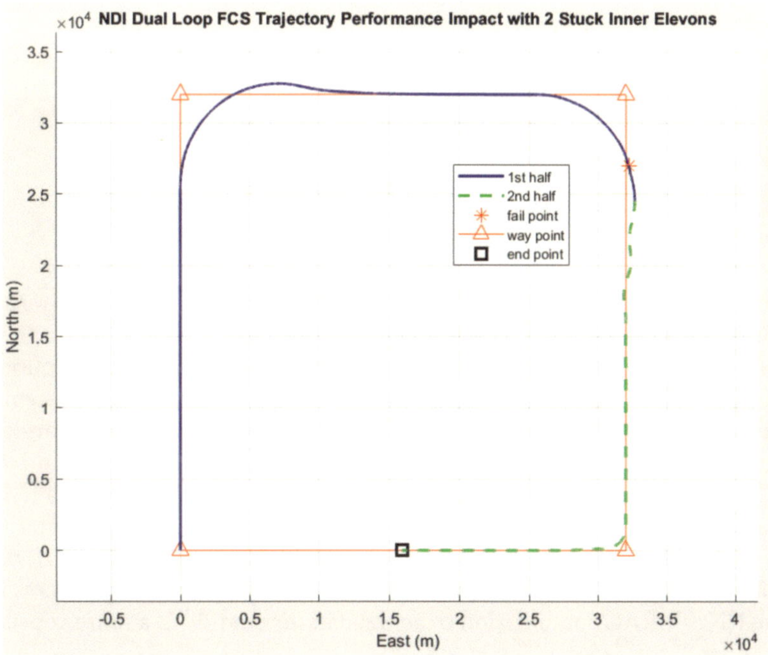

Figure 8.
ADMIRE flight trajectory performance degradation with two stuck IB Elevons (Mission ended prematurely with attitude pointing accuracy severely degraded).

4. IFC algorithm description

IFC algorithms have become popular in improving FCS performance, especially when employing the Optimal Control Modification (OCM) [11] as an adaptive law design modification, to further enhance the controller's ability to cope with unknown

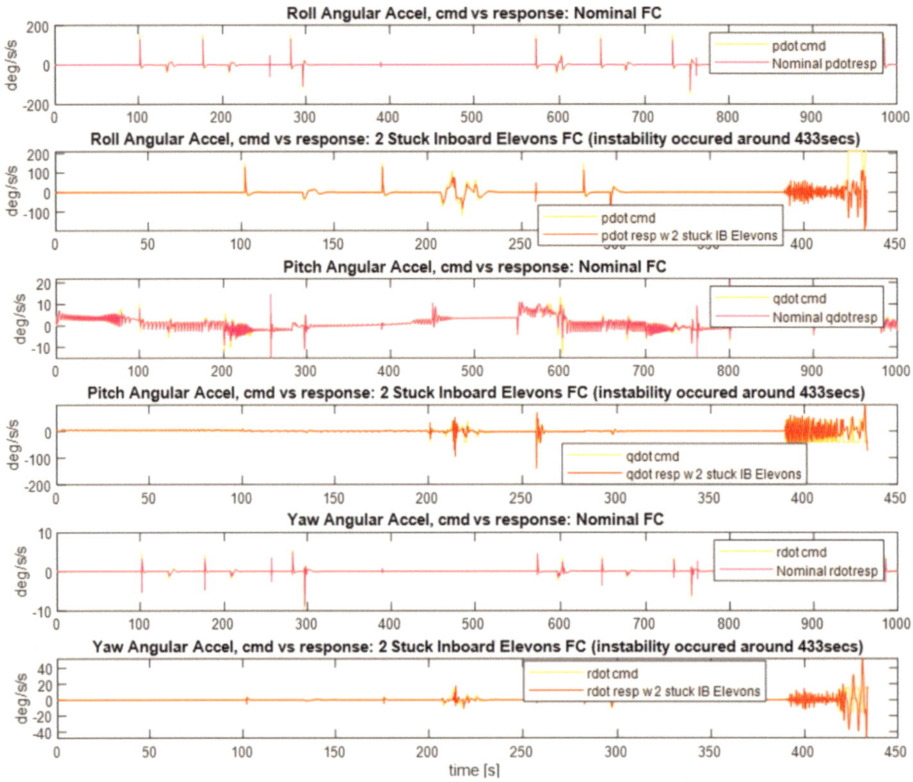

Figure 9.
FCS angular acceleration command following via NDI – Nominal vs. 2 stuck IB Elevons.

uncertainties, aircraft damage, and disturbance attenuation since the early 2000s. We have teamed with NASA Ames Research Center to leverage their work on the IFC development for the NASA GTM piloted simulation [12], and the actual piloted F-15 and F-18 [13] flight tests of the OCM adaptive control law at NASA Armstrong Flight Research Center [4]. Toward this end, we have adopted their GTM-IFC design for robustness enhancement of the ADMIRE FCS when subjected to degraded control surfaces. The IFC block that we adopted from [4, 14] can be considered a hybrid design, as shown in **Figure 12**. The baseline adaptive control block (shown in yellow) is architected using the direct Model Reference Adaptive Control (MRAC), while the Neural Network Controller (NNC) block offers an adaptive learning process that allows the NNC-MRAC combination to achieve its adaptation in a more consistent and robust manner.

The IFC architecture developed for GTM [4] that is adopted here is shown in **Figure 13**. For the ADMIRE FCS performance enhancement subject to imperfect cancelation of the NDI CLAW due to off-nominal flight conditions (i.e., stuck control surfaces and/or complete loss of some control surfaces). For the sake of describing how the IFC design developed in [4] for aircraft flight control is applied to the ADMIRE's robustness enhancement study, we re-use Section B of [4] (i.e., neural network (NN) direct adaptive control) and define its input/output (I/O) for the neural network adaptive control signal, u_{ad}, and how that signal is computed using the ADMIRE aircraft states vector.

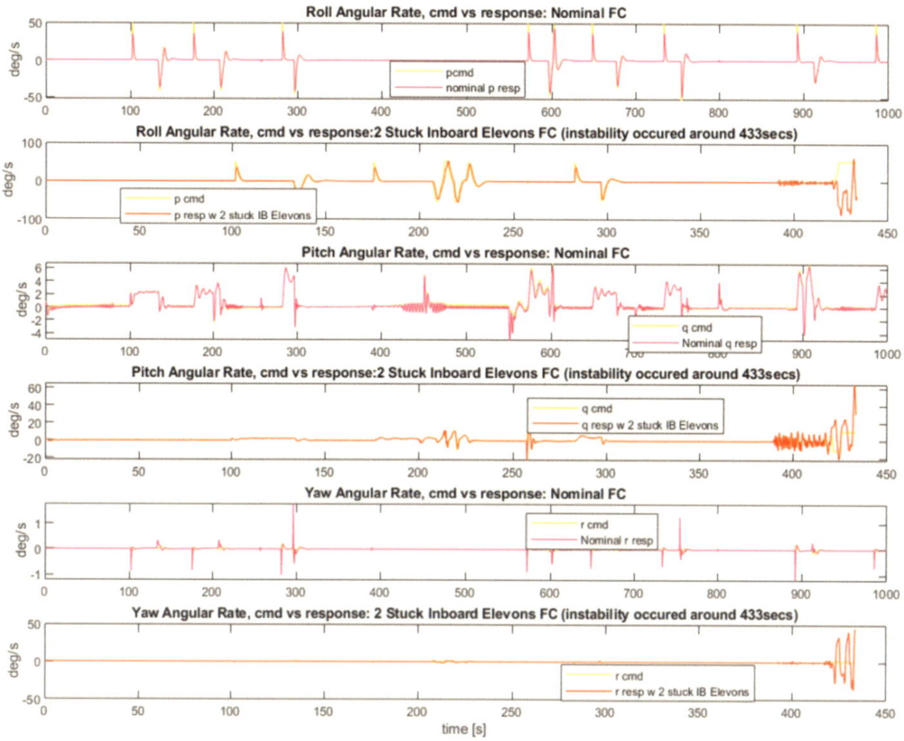

Figure 10.
FCS angular rate command following via NDI – Nominal vs. 2 stuck IB Elevons.

4.1 NN-based adaptive control using conventional MRAC

The adaptive control augmentation vector u_{ad} is based on the adaptation law by Rysdyk and Calise [15] with a modification to include additional product terms that appear in the nonlinear plant dynamics described by Eqs. (31)–(33) of [4]:

$$u_{ad} = W^T \beta_{nn}(C_1, C_2, C_3, C_4, C_5, C_6) \tag{2}$$

where β_{nn} is a vector of basis functions computed using Kronecker products with Ci, $i = 1; \ldots ; 6$, as inputs into the neural network consisting of control commands, sensor feedback, and bias terms. More specifically, the product terms are

$$C_1 = V^2 \left[\omega^T \; \alpha\omega^T \; \beta\omega^T\right] \tag{3}$$

$$C_2 = V^2 \left[1 \; \alpha \; \beta \; \alpha^2 \; \beta^2 \; \alpha \; \alpha\beta^2\right] \tag{4}$$

$$C_4 = \left[p\omega^T \; q\omega^T \; r\omega^T\right] \tag{5}$$

$$C_5 = \left[u\omega^T \; v\omega^T \; w\omega^T\right] \tag{6}$$

$$C_6 = [1 \; \theta \; \phi \; C_T] \tag{7}$$

The NN basis function, β_{nn}, then expressed as

$$\beta_{nn} = [C_1 \;\; C_2 \;\; C_3 \;\; C_4 \; C_5 \; C_6]^T \tag{8}$$

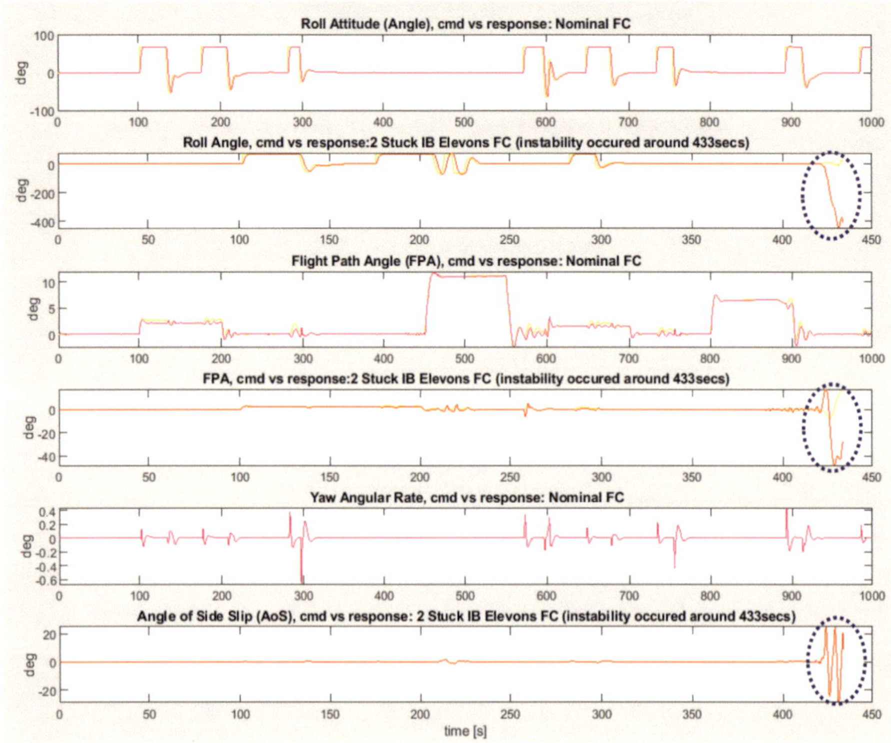

Figure 11.
FCS Bank angle, flight path angle (FPA), and AoS command following via NDI – Nominal vs. 2 stuck IB Elevons (causing instability occurred at 433 secs).

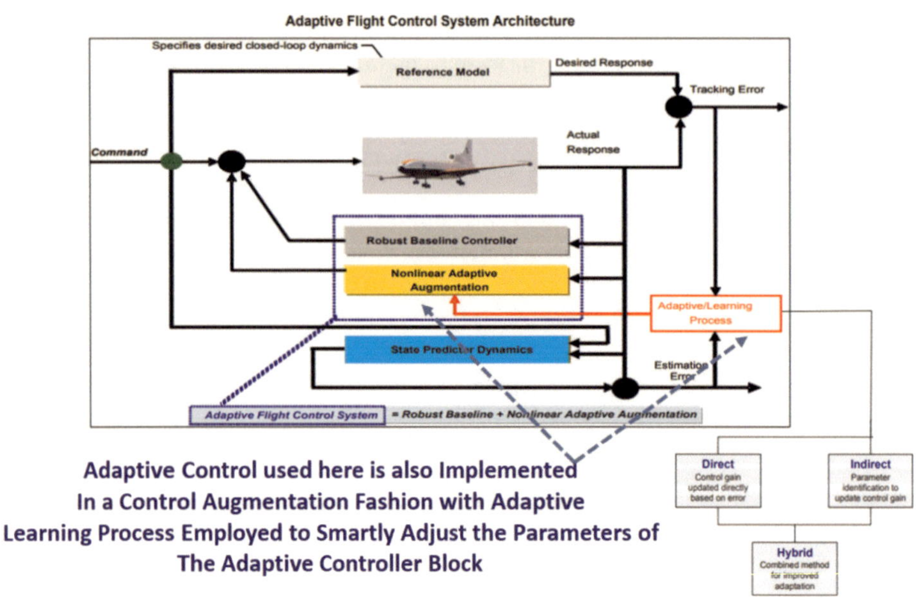

Figure 12.
Adaptive control motivation for why a hybrid design should be exploited to achieve verification and validation [14].

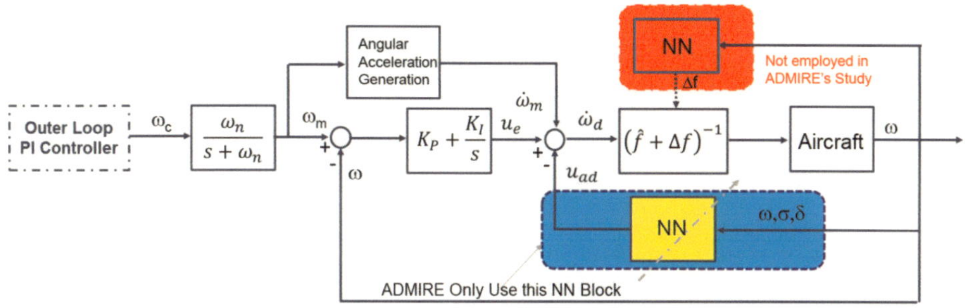

Figure 13.
NNC architecture (yellow block) and their I/O Interface with ADMIRE Aircraft's state vector.

The network weights W are computed by a direct adaptive law, which incorporates a learning rate G > 0 and an e-modification term [14] μ >0 according to the following weight update law

$$\dot{W} = -\Gamma\left(\beta_{nn}e^{T}PB + \mu\left|\left|e^{T}PB\right|\right|W\right) \tag{9}$$

where Γ is an adaptation gain matrix, the matrix P solves the Lyapunov equation $A^{T}P + P^{T}A = -Q$ for some positive-definite matrix Q, e is the model following error, and ||.|| is a Frobenius norm. **Table 1** shows how P is computed for use in the ADMIRE IFC application. The β_{nn} implementation (of Eq. (8)) is shown in **Figure 14**.

The e-modification term in Eq. (9) provides robustness in the adaptive law [16, 17]. The weight update law in Eq. (9) guarantees the stability of the neural network weights and the tracking error. The proof of this updated law using the Lyapunov method is provided by Rysdyk and Calise [18].

In the above expressions, [α, β, ϕ] are the AoA, AoS, and bank angle, [u v w] are the velocity vector components in the body frame, and ω is the vehicle body frame angular velocity vector.

5. Derivative-free model reference adaptive control (DF-MRAC)

The DF-MRAC design framework is fully described in the recently published book written by Yucelen and Calise [19]. Here, we just want to describe how it is being used

```
%% See the IFC/NN Based Controller Block
ifc.Kp_p = 3; ifc.Kp_q = 16; ifc.Kp_r = 5;
Ap = −ifc.Kp_p; Bp = ifc.Kp_p; Cp = 1; Dp = 0;
Aq = −ifc.Kp_q; Bq = ifc.Kp_q; Cq = 1; Dq = 0;
Ar = −ifc.Kp_r; Br = ifc.Kp_r; Cr = 1; Dr. = 0;
Kp = diag([ifc.Kp_p,ifc.Kp_q,ifc.Kp_r]);
Pmat = lyap(-Kp',eye(3));
Pmat = [0.1667 0 0
0 0.0312 0
0 0 0.1000];
Bmat = eye(3);
```

Table 1.
Lyapunov function block used in ADMIRE IFC block.

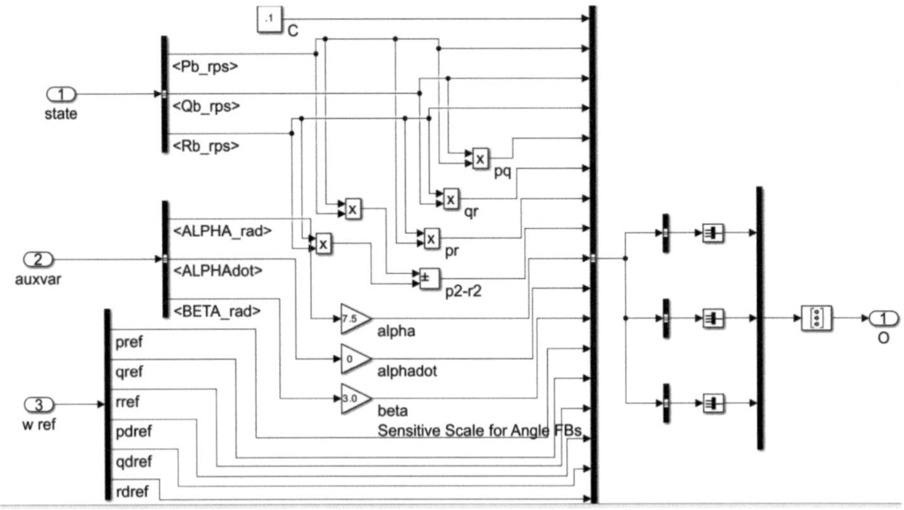

Figure 14.
β_{nn} implementation in ADMIRE FCS IFC block (not full state FB).

to solve the problem of two stuck inboard elevons by mixing its solution with the IFC architecture described in the previous section. The adaptive law in DF-MRAC has the form

$$\hat{W}(t) = \Omega_1 \hat{W}(t-\tau) + \hat{\Omega}_2(t), t > \tau \tag{10}$$

where $\tau > 0$, Ω_1 and $\hat{\Omega}_2(t)$ satisfy:

$$0 \leq \Omega_1^T \Omega_1 < I_s \tag{11}$$

$$\hat{\Omega}_2(t) = \kappa_2 \beta(x(t)) e^T(t) PB, \kappa_2 > 0 \tag{12}$$

There is no need for e-modification as in (9) with this adaptive law, and the time delay τ can be freely chosen. DF-MRAC has many other advantages not present in MRAC that are illustrated in [19], including a natural robustness to unmodeled dynamics, greatly improved performance when augmenting a baseline controller that employs proportional + integral control, and the ability to treat uncertainties characterized by time-varying ideal weights. Conventional MRAC employs the assumption that the uncertainty must be characterized by a set of constant ideal weights, whereas with DF-MRAC, the ideal weights can be time-varying. This greatly reduces the burden on the designer to carefully select the correct set of basis functions in the design process. Another added advantage is that the time delay parameter employed in DF-MRAC adds an additional degree of freedom that has the added advantage of introducing greater memory into the learning process.

Figures 15 and **16** illustrate how DF-MRAC has added an option to the ADMIRE simulation. To date, it has been found that this option greatly improves the ability to handle failure in actuation (**Figure 17**). The ADMIRE aircraft is now able to restore its stabilization and maintain its desired mission performance for all the alternative CA methods that are currently implemented. This improved performance is illustrated in **Figures 18, 19–22** for the same case of stuck inboard elevons, previously shown in **Figures 6**, 7 and **8–11** without adaptation. **Figures 18** and **19** present the IFC's ability to

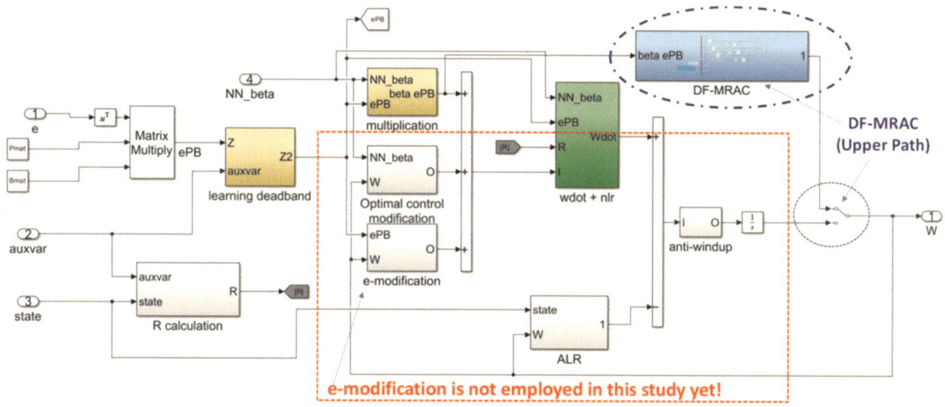

Figure 15.
DF-MRAC implementation in parallel with derivative-based MRAC.

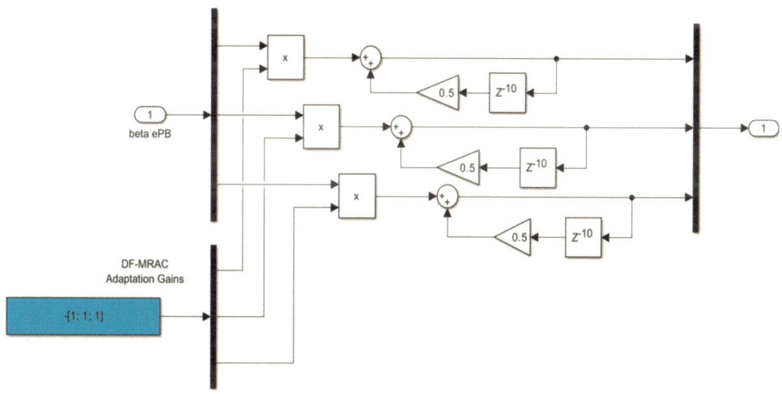

Figure 16.
DF-MRAC implementation within the IFC architecture.

maintain the aircraft controllability and restore stability when subjected to two stuck IB elevons and maintain a good command following this degraded flight condition (**Figure 19**).

6. IFC extension with direct adaptive control (DAC) algorithm to CA interaction

This section is a preliminary effort addressing F/M closed-loop regulation (i.e., enhancing F/M response to F/M commands generated by the inner loop NDI controller), which is the immediate layer tightly connected and interacting with the CA functional block presented in **Figure 23**. This is still being viewed as an adaptation protection layer which is implemented on the CLAW side (and not the CA side). This is a part of the hybrid direct and indirect adaptive control design framework presented earlier in **Figure 12**.

The closest connection to CA in an adaptive sense (without explicitly solving the dynamic CA problem as described in [20–22]) is to actively regulate the angular acceleration error with some quadratic minimization adaptive regulator. We chose the

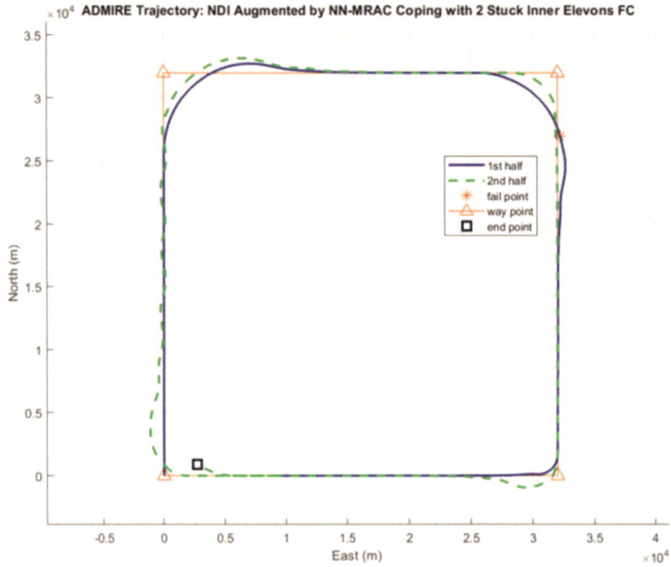

Figure 17.
IFC/NDI restoring stabilization and control subject to stuck inboard Elevons sufficient to complete the Mission by employing DF-MRAC.

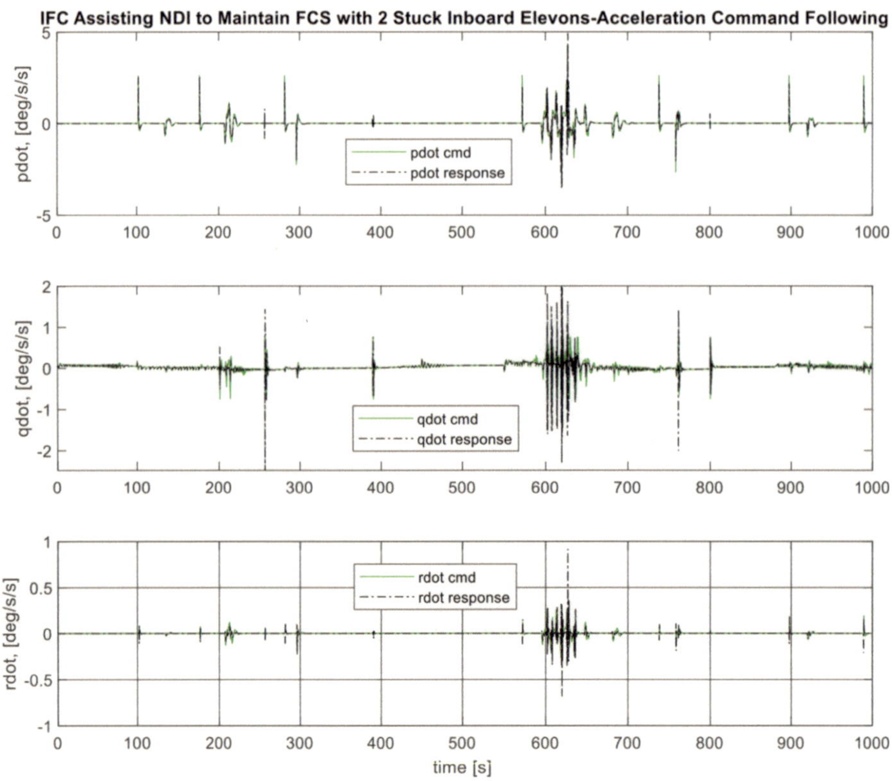

Figure 18.
Restored stabilization by IFC with improved angular acceleration command following subject to 2 stuck inboard Elevons (see zoom-in region in ***Figure 19*** *for command following accuracy).*

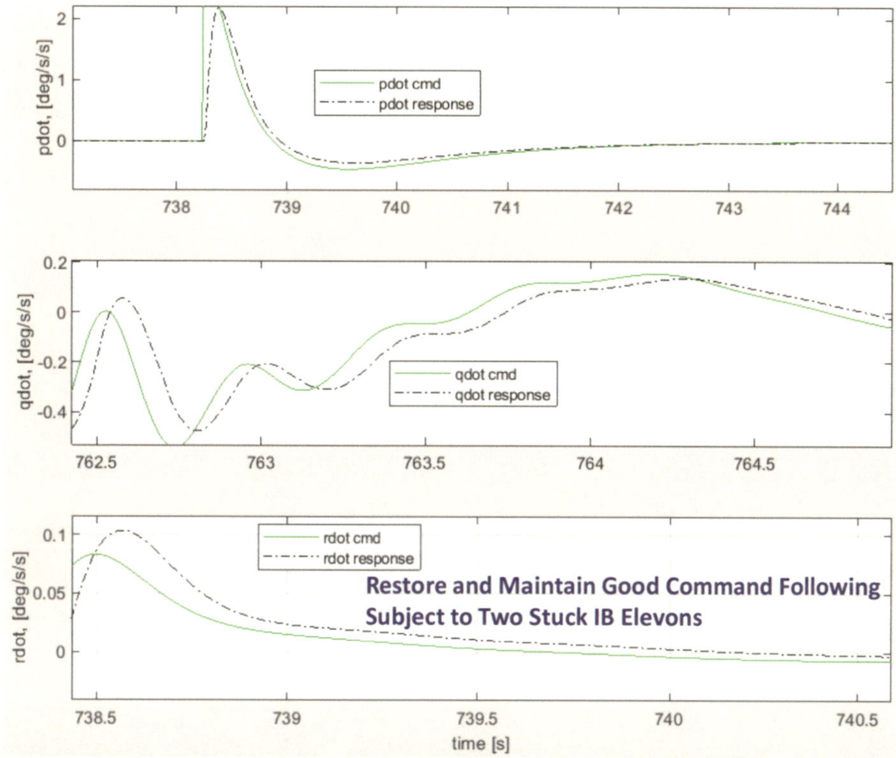

Figure 19.
Restored stabilization by IFC with a zoom-in snapshot for command following restoration illustration (presented in ***Figure 18****).*

direct adaptive control algorithm presented in [10] (and recently demonstrated in [9]) to actively enforce angular acceleration commands in the presence of degraded effectors. Surprisingly, the use of such a direct adaptive control scheme as an adaptive acceleration error regulator works quite well for both stuck control surfaces and degraded effector's deflection angle and deflection rate limits (see **Figure 22** for its adaptive gain real-time adjusting as a function of model following errors of angular acceleration vector).

The CA design framework has been employed by the aircraft, UAVs, and spacecraft industries for more than five decades. However, various design algorithms and their design evolution are only captured by technical publications and internal technical memorandum until 2017 with a formal textbook by Durham et al. [6]. This textbook is viewed as a compilation of multi-decade research and development (R&D) in CA algorithms for aircraft FCS, especially with the mixing of NDI-based design and the Cascaded Generalized Inverse (CGI) for X-35 and F-35 applications [8]. Readers are referred to Chapters 7 and 8 of the textbook [6] for the formal CA design framework captured therein and CA applications to the X-35 in Chapter 9. The concepts of *Desired Commands* (for Commanded Moments Generation by the inner loop NDI CLAW) vs. *Attenable Moments Set (AMS)* via *Admissible Control Effectors* captured in Appendix A of [6] and how they are being connected to the aircraft FCS design is captured in Appendix B. Note that the CA framework should be generalized for 6DOF control with full 6DOF F/M command following as described in **Figure 22** rather than solely moment or torque regulation.

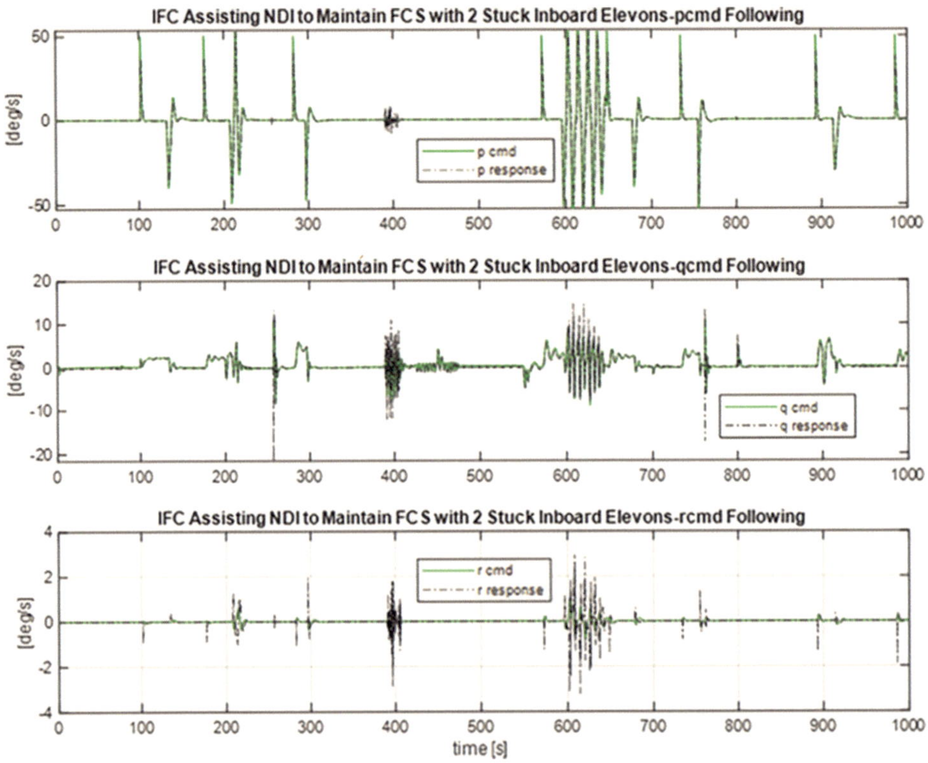

Figure 20.
Restored stabilization by IFC with improved angular rate command following subject to 2 stuck inboard Elevons.

The CA design framework interconnected to the CLAW side has been an active research topic during the last 10 years (e.g., innovative control effectors (ICE) [23] and balancing/resolving actuators redundancy via control vs. control allocation [24]). It has become an important topic for maintaining high-speed UAV missions, especially under stressful operating conditions that include degraded effectors due to thermal impact or damages resulting from adversarial actions (see [25, 26]).

The primary goal of adaptive CA dealing with control effector failures is to dynamically reshape the CA multiple effectors blending matrix B (in **Figure 22**) to explicitly zero out the corresponding column of such a failed effector so that the inner loop NDI CLAW does not rely on the same original number of healthy effectors for F/M contribution. In this example case, the control effector will be the second column, as illustrated in **Figure 23**. Therefore, the new B matrix used by the CLAW will now have a new dimension of [3x6] matrix (restructuring the B matrix by removing the second column of its original [3x7] matrix) if the second effector has been detected and declared as a complete loss of its operational capacity. Of course, other CA algorithms could offer effectors fault tolerant CA capabilities without explicitly restructuring the mixing matrix B such as in [27].

7. Concluding remarks and future directions

The study presented herein, adaptive control augmentation via IFC to assist the primary baseline FCS subject to off-nominal FCs, is not new (e.g., see [4] and its cited

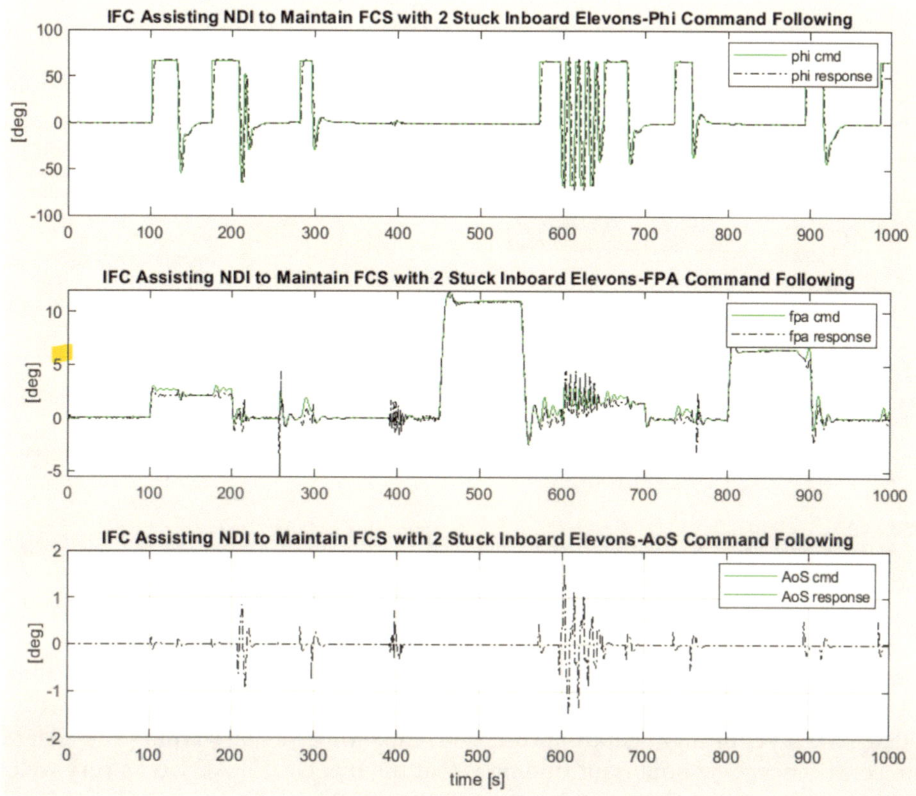

Figure 21.
Restored stabilization by IFC with improved angle command following and tracking subject to 2 stuck inboard Elevons.

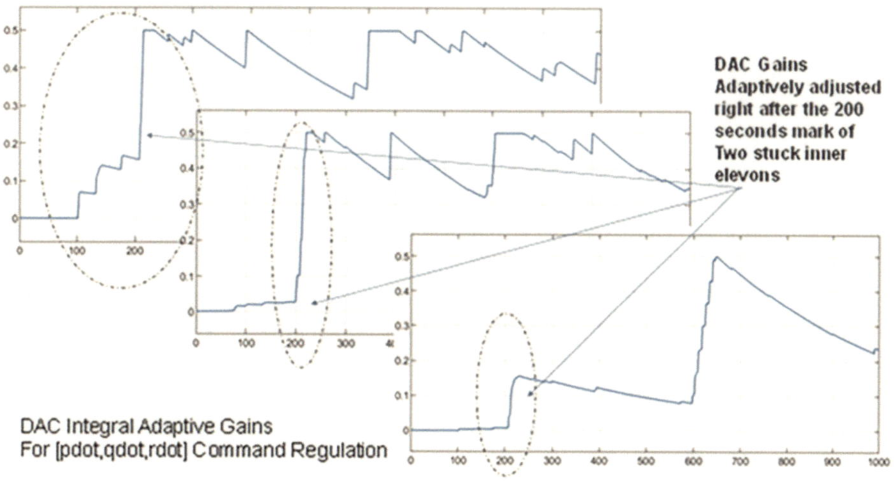

Figure 22.
DAC gain adaptation coping with two stuck IB Elevons.

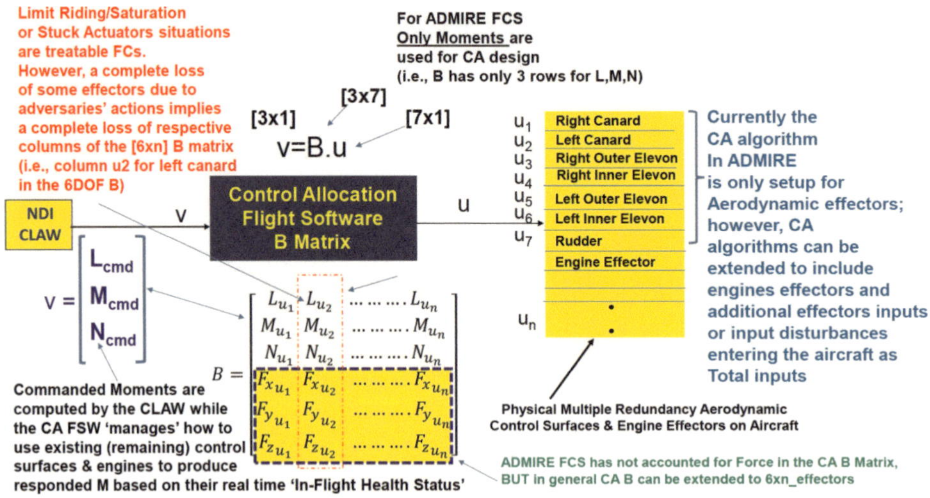

Figure 23.
Full F/M regulation with CA framework design coupled with CLAW where 3DOF force has been added to the baseline moments.

references therein or [14] for a formality of adaptive control verification and validation). We re-use such an attractive IFC solution for high-speed UAV applications as an added extra layer of flight control protection, exploiting its self-learning and adaptation consistency. One important finding is that the IFC DF-MRAC works very well with all existing seven CA algorithms to maintain FCS performance, while the DAC algorithms are only operational to three of them using the same set of adaptation parameters selection. This means the DAC algorithm requires retuning of its adaptation parameters ($[\Gamma_I, \Gamma_p, \sigma]$, see [10]) for different CA algorithms, while the IFC algorithms do not require retuning. The current IFC design will be continuously evaluated, revised, and upgraded to address more formal failure cases and to mature its design to serve more realistic high-speed UAV missions.

Future directions will extend the IFC design to address the functional CA performance aspects (augmentation of the CA functional blocks subject to effectors performance degradation, e.g., see [20–22, 25, 26] or [27]) rather than just adaptation augmentation added on in the control law side of the FCS. As mentioned earlier in Section 5, there are two main paths (e.g., see [24]) to dynamically size/shape and determine the commanded effectors: (1) directly determine the real deflection of effectors from the applied optimal control and (2) use optimal control or NDI to compute the total F/M, and then use the actual CA function (via the multiple effectors mixing matrix B in **Figure 23**) to compute individual effectors' deflection in real-time. Aircraft CA algorithms have been extensively studied during the past two decades (e.g., see [6] and references cited therein) and are still considered an active research area in addressing CA mixing to achieve (i) optimal performance when all effectors are healthy and (ii) suboptimal performance when effectors are in degraded operating conditions (e.g., see [28] for actively reshaping the effector's blender solution B matrix in real-time (see **Figures 4** and 5 of [28])) while still able to maintain the designated mission.

Future improvements are contemplated for IFC in general and DF-MRAC in particular. These include:

a. Adaptive hedging to improve response when actuator position and rate limits are active and to account for the fact that actuators are bandwidth limited (see [29]).

b. Adaptive loop transfer recovery to guarantee the gain, phase, and time delay margins of the baseline design are preserved under IFC (see [30]).

c. A direct method for adaptation to actuator failures that do not require augmentation of the CA function blocs. This involves a modification to Eq. (11) as described in Section 3.4 of [19].

d. Application of a neural network-based control to attenuate external disturbances developed by Levin and Ioannu in [31].

e. Possibility of employing the multiple model adaptive mixing schemes developed in [32, 33] to ensure a wider region of adaptation without model switching.

Acknowledgements

This study is part of a bigger campaign in preparation for future high-speed UAV business pursuit and was partially funded by the Jesplsoft IRAD Program. The NASA partnership forming and support are greatly appreciated.

Conflict of interest

The authors declare no conflict of interest.

Author details

Quang M. Lam[1*], Anthony J. Calise[1] and Nhan T. Nguyen[2]

1 JESPLSOFT Corp, Fairfax, USA

2 NASA Ames Research Center, Moffett Field, USA

*Address all correspondence to: quang.lam@jesplsoft.com

References

[1] Eva Wu N. Reliability analysis for AFTI-F16 SRFCS using ASSIST and SURE. In: Proceedings of the American Control Conference Anchorage, AK May 8-10, 2002

[2] Khan R, Williams P, Riseborough P, Rao A, Hill R. Active fault tolerant flight control system design - A UAV case study. arXiv:1610.03162v1. 2016

[3] Hu B, Seiler P. A probabilistic method for certification of analytically redundant systems. IEEE. 2013

[4] Nguyen N, Krishnakumar K, Kaneshige J, Nespeca P. Flight dynamics and hybrid adaptive control of damaged aircraft. Journal of Guidance, Control, and Dynamics. 2008

[5] Forssell L, Nilsson U. ADMIRE The Aero Data Model in Research Environment Version 4, Model Description, Report No. FOI-R–1624–SE, FOI Swedish Defense Research Agency, Systems Technology, SE –164 90, Stockholm. 2005

[6] Durham W, Bordignon K, Beck R. Aircraft Control Allocation. Wiley Aerospace Series; 2017

[7] Lathasree P, Pashilkar AA, Sundararajan N. Application of generic flight controller design approach for a Delta canard fighter aircraft-ADMIRE. In: Sixth Indian Control Conference (ICC) December 18–20, 2019. India: IIT Hyderabad; 2019

[8] Bordignon K, Bessolo J. 'Control allocation for the X-35B,' AIAA 2002-6020. In: 2002 Biennial International Powered Lift Conference and Exhibit, 5–7 November 2002. Virginia: Williamsburg

[9] Lam Q. Direct Adaptive Control Augmentation and Dynamic Control Allocation algorithms Subject to Stuck Control Elevons for Aircraft Flight Control Robustness and Performance Enhancement, Jesplsoft IRAD Report. 2023. Available from: https:// Jesplsoft.com

[10] Lam Q, Barkana I. Direct Adaptive Control Treatment to Flight Control Input Saturation. AIAA; 2005

[11] Nguyen N. Model-Reference Adaptive Control - A Primer. Springer-Verlag; 2018. ISBN: 3319563920

[12] Campbell S, Kaneshige J, Nguyen N, Krishnakumar K. An adaptive control simulation study using pilot handling qualities evaluations. In: AIAA Guidance, Navigation, and Control Conference, AIAA-2010-8013, August 2010

[13] Nguyen N, Hanson C, Burken J, Schaefer J. Normalized optimal control modification and flight experiments on NASA F/A-18 aircraft. AIAA Journal of Guidance, Control, and Dynamics. 2017; **40**:1061-1075

[14] Nguyen N. Verification and Validation Challenges for Adaptive Flight Control of Complex Autonomous Systems. AIAA; 2018

[15] Li X, Liu S, Guo F, Zhang S. Fault-tolerant control based on control allocation for hypersonic vehicle with actuator stuck fault. IEEE. 2019

[16] Narendra KS, Annaswamy AM. New adaptive law for robust adaptation without persistent excitation. IEEE Transactions on Automatic Control. 1987;**32**(2):134-145

[17] Calise AJ, Sharma M, Corban JE. Adaptive autopilot Design for Guided

Munitions. Journal of G&C Dynamics. 2000

[18] Rysdyk RT, Calise AJ. Fault tolerant flight control via adaptive neural network augmentation. In: AIAA Guidance, Navigation, and Control Conference, AIAA 1998–4483. 1998

[19] Yucelen T, Calise AJ. Derivative-free adaptive control. AIAA. 2023;**267**

[20] Lu B, Ma J, Zheng Z. Adaptive closed-loop control allocation based fault tolerant flight control for an over-actuated aircraft. IEEE. 2019

[21] Jingping S, Xi LVY, Xiaobo Q, Jing S. A coordinated control method of thrust vector and aerodynamic surfaces based on control allocation technology. IEEE. 2018

[22] Cui L, Zuo Z, Yang Y. A control-theoretic study on iterative solution to control allocation for over-actuated aircraft. IEEE. 2018

[23] Niestroy MA, Dorsett KM, Markstein K. A Tailless Fighter Aircraft Model for Control-Related Research and Development. AIAA; 2016

[24] Häarkegård O, Glad T. Resolving Actuator Redundancy: Optimal Control vs. Control Allocation. 2004. Available from: http://www.control.isy.liu.se

[25] Yong F, Ji-Hong Z, Jia-Qiang Z, Zeng-Qi S. Genetic Algorithm Based Constrained Control Allocation for Tailless Fighter. The National High Technology Research and Development Program of China Grant #2005AA751010

[26] Yang L-Y, Zhong Y-W, Shen G-Z. Control allocation and Management for Aircraft with multiple effectors. IEEE. 2009

[27] Hamayun MT, Edwards C, Alwi H, Bajodah A. A fault tolerant direct control allocation scheme with integral sliding modes. International Journal of Applied Mathematics and Computer Science. 2015;**25**(1):93-10

[28] Harris JJ, Stanford JR. F-35 Flight Control Law Design, Development, and Verification. AIAA Aviation Forum; 2018

[29] Johnson EN, Calise A. Limited authority adaptive flight control for reusable launch vehicles. Journal of Guidance, Control and Dynamics. 2003; **26**(6):926-913

[30] Calise AJ, Yucelen T. Adaptive loop transfer recovery. AIAA Journal of Guidance, Control, and Dynamics. 2012; **35**(3):807-815

[31] Levin J, Ioannu PA. Adaptive control with neuro-adaptive disturbance rejection. In: 17th Mediterranean Conference on Control & Automation Makedonia Palace. Thessaloniki, Greece; 2009

[32] Kuipers M, Ioannou P, Fidan B, Mirmirani M. Analysis of an adaptive mixing control scheme for an Airbreathing hypersonic vehicle model. In: American Control Conference Hyatt Regency Riverfront; June 10–12, 2009; St. Louis, MO, USA

[33] Lam Q, Cloutier J, Hart A, Stockbridge S. Exploring multiple model control mixing for robustness and performance enhancement of hypersonic vehicles. In: Proceedings of the Joint Army Navy-NASA-Air Force Conference, December 2014

Chapter 7

Practical Approaches to the Control of Milk Fermentation with Kefir Grains

Jožef Ritonja, Andreja Goršek, Darja Pečar, Darius Andriukaitis and Boštjan Polajžer

Abstract

In the chapter, milk fermentation for kefir production is studied. The traditional kefir production process based on inoculating kefir grains into milk is considered. The quality and quantity of the produced kefir also depend on the dynamics of the fermentation process. The chapter presents the design and synthesis of the closed-loop control system in which changing the bioreactor's temperature is used to control the time course of the concentration of dissolved CO_2. In the chapter: (1) a nonlinear dynamic mathematical model of the fermentation process, which allows evaluating the influence of the bioreactor's temperature on the dynamics of the fermentation process, is presented; (2) the design and synthesis of a conventional linear control system with constant parameters are carried out; (3) an adaptive control system that enables the tracking of the courses of the quantities of the fermentation process to the desired reference trajectories without the time-consuming preliminary identification of the parameters of the fermentation process model is developed. The numerical, experimental, and analytic outcomes of the study are presented.

Keywords: milk fermentation, batch bioreactor, mathematical modeling, control system design and synthesis, linear control, model reference adaptive control

1. Introduction

In the chapter, milk fermentation for kefir production is studied. The traditional kefir production process based on inoculating kefir grains into milk is considered. Characteristics, quality, and quantity of the produced kefir depend on the used microorganisms and substrate, but they also depend on the time course of the fermentation product concentration [1]. Although different types of bioreactors can be used for kefir production, batch bioreactors are highly practiced [2]. The specific operation of batch bioreactors allows their simple construction, which is returned with easy maintenance and low production costs of bioreactors [3]. Therefore, batch bioreactors are the most widely used bioreactors. Unfortunately, in batch bioreactors, it is not possible to add or remove individual substances during operation and thus influence the course of the fermentation process [4]. This represents a major limitation in the use of existing batch bioreactors. Fermentation processes in the existing industrial and laboratory batch bioreactors run independently (autonomously). The courses depend only on the initial concentrations of substances, and once the fermentation processes have been started, it is no longer possible to influence the execution of the fermentation processes. This problem is well-known in the industrial and academic environment. There are many publications showing this problem [2–4] and possible solutions [5–7].

This chapter presents another option to control the time course of the fermentation process. The possibility of controlling the time course of the fermentation process by changing the bioreactor's temperature instead of adding individual substances during the operation is investigated in the chapter thoroughly.

In the study's initial phase, we also reviewed relevant journal publications considering fermentation process control published in the period 2000–2020. The purpose of the review was to determine the direction of research in this area and to analyze the existing solutions to this problem. We used established research databases in both related fields, that is, Electrical Engineering/Control and Bioprocess Engineering/Bioreactors. As expected, we found a lot of recent interesting work in the field of Continuous Bioreactor Control and in the field of fed-batch bioreactor control. Some interesting articles concerning the use of advanced control techniques in continuous bioreactors are presented in Refs. [5–11]. The implementation of advanced control methods for fed-batch bioreactors can be found in many articles; some of them are [12–15]. Unexpected and very interesting, however, was the disclosure that there are very few publications dealing with the control of batch bioreactors. We found only a few papers where the control of the batch bioreactors was realized. In both cases, the manipulation of airflow of oxygen concentration was used for control. In Ref. [16], gain scheduling control was used, and a conventional PI control strategy was proposed in Ref. [17]. This was our additional motivation to work more in-depth and intensively in the field of control of fermentation processes in batch bioreactors.

The problem of control of fermentation processes in batch bioreactors is frequent and up-to-date. Based on the increasing interest in a suitable solution and on the basis of already existing knowledge presented in publications, we started with the multidisciplinary multiannual cooperation of three faculties. Our work was divided into three stages: (1) design of the control system, which will enable assure the prescribed operation in batch bioreactors; (2) derivation of the mathematical model of the fermentation process in batch bioreactors which also describes the impact of operating conditions on the execution of the fermentation process and will be suitable for the analysis and simulation of the fermentation process and for the development of the control system; (3) synthesis of the advanced control system, which will approve a noncomplicated and practical implementation. The work done has been addressed in some publications. The use of the heating system for control of the fermentation process in batch bioreactors was presented in Refs. [18–22], and use of the stirrer system in Refs. [18, 23]. In this chapter, the continuation of work in the field of kefir production control is presented and the findings are summarized and collected.

The basic hypothesis of our work was that the fermentation process in batch bioreactors can be controlled by changing the temperature in the bioreactor. In this way, it is possible to achieve the desired time courses of concentrations of substances in the fermentation process and thus influence the quality and quantity of the fermentation product. To achieve this goal, we need a control system, which must be simple enough to be implemented in a batch bioreactor. It is also necessary that the tuning of the controller parameters is automatic and that the users of the bioreactors will not need additional knowledge and time-consuming setting of the parameters of the control system. The second hypothesis was that a control system that will meet the set requirements can be realized based on the theory of adaptive systems.

For this purpose, the chapter presents a nonlinear dynamic mathematical model of the fermentation process, which allows evaluation of the influence of operating parameters (stirrer speed, temperature) on the dynamics of the fermentation process. Based on this model, the design and synthesis of a conventional linear control system with constant parameters can be carried out. Accurate knowledge of the fermentation process model parameters is required for tuning the parameters of a conventional linear controller, which significantly reduces the usability of such a control system. The chapter's emphasis is on the development of an adaptive control system that enables the following of the responses of the quantities of the fermentation process to the desired reference trajectories without the time-consuming preliminary identification of the parameters of the fermentation process model. The main contribution of the chapter is the presentation of the development of a control system that will enable the control of the fermentation process for kefir production even in uncomplicated batch bioreactors. The developed control system confirmed both hypotheses.

The chapter is organized into nine Sections. In Section 2, a considered milk fermentation process with kefir grains is described. In Section 3, the available laboratory system consisting of a batch bioreactor with a controlled heating/cooling system, measurement equipment, and equipment for data acquisition and control is presented. In Section 4, a mathematical model of the fermentation process in the batch bioreactor is described. Results obtained with the derived mathematical model are shown in Section 5. A batch bioreactor's closed-loop control system is developed based on the knowledge gained from the mathematical model analysis. The design and synthesis of the control system are carried out in two ways: By using the conventional linear control theory and on the basis of the advanced adaptive control theory, both concepts are shown in Section 6. The results of both control approaches are presented in Section 7. Discussion, comments, and concluding remarks are offered in Sections 8 and 9.

2. Controlled fermentation process

The given study deals with milk fermentation. The release of CO_2 is studied, and the focus is on fermentation with kefir grains. Inoculum-initiated milk fermentation lasts approximately one day. During this time, homofermentative lactic acid streptococci grow fast, at the beginning producing a decrease in pH. Reduced pH causes a decrease in the number of streptococcus and at the same time accelerates the growth of lactobacilli. Yeasts and increased temperature have a positive effect on the growth of heterofermentative streptococci. These have a significant influence on the aroma of the product.

Before the considered fermentation process, kefir grains were activated for five consecutive days (daily washing with cold water and after that putting in milk at room temperature). For starting of the fermentation process the pasteurized fat milk was preheated and inoculated with activated grains. By using different kefir grains, the fermentation processes with different dynamic and static characteristics were performed. The considered fermentation processes are nonlinear dynamic processes that require sophisticated control algorithms to achieve desired behavior and consequently quality fermentation products.

3. Used lab apparatus

A laboratory batch bioreactor with additional equipment was applied to show the applicability of the proposed practical approach for the control of the kefir fermentation process. This laboratory system was also used for the thorough analysis and modeling of the fermentation process. Before the implementation of the control system, a systematic analysis through the set of experiments was carried out. First, a set of experiments with constant operating conditions was performed. Then, a multitude of tests with the changeable temperature in the bioreactor was done. The goal of these experiments was to discover the impact of the temperature of the fermentation substances on the dynamics of the fermentation process. During these experiments, the input quantity of the analyzed fermentation process was the temperature in the bioreactor, and the output quantity was the dissolved CO_2, which represents the controlled fermentation product.

3.1 Batch bioreactor

Fermentation processes were performed in the reaction calorimeter Mettler Toledo RC1e, which was applied as a batch bioreactor. A comprehensive explanation of RC1e is presented in Ref. [22].

3.2 Integrated system for heating and cooling of fermentation mixture

The tested bioreactor was originally provided with the system for heating and cooling (heating/cooling system). The silicone oil was implemented as a transfer liquid, which is injected into a closed circulation system through the double jacket of the reactor. The heating/cooling system is completed with the closed-loop temperature control system. A proportional-integral controller was used. The heating/cooling system enables the adjustment of the bioreactor's temperature in the area from 5 to 50° C. Due to the controller's integral characteristic, the control system operates without steady-state error for a constant reference temperature. The lag in the control system is substantial in comparison to the lag of the fermentation process. The heating/cooling system was modeled with the 1st order lag term with unit gain and time constant $T_{\vartheta cs}$ = 0.1 h.

3.3 Dissolved CO_2 measurement

The choice of the controlled plant output quantity is decisive for the selection of the control approach and for the practical realization. The selected quantity must comprehend the essential information about the fermentation process. It is also required the unproblematic, accurate, and fast realization of its measurement.

For the control and monitoring of the fermentation processes, the measurements of dissolved oxygen and cell culture are fundamental. Oxygen concentration is very important because it correlates with growth and metabolite. Another quantity important in the execution of the fermentation process is the concentration of the biomass.

During laboratory experiments or operating in an industrial environment, it is very difficult to measure dissolved oxygen and/or cell culture concentrations. Therefore, in the presented contribution, we have shown how we can use the measurement of dissolved CO_2 concentration instead of measuring these two quantities [24]. CO_2 is the outcome of microorganism metabolism and represents a good indicator of the course of fermentation. A good feature of using CO_2 measurement as an indicator of the fermentation process is the ease of CO_2 measurement. CO_2 is homogeneously distributed in the reactor mixture, the sensors are accurate, reliable, have a long lifespan, their measurement curve is known [25]. The CO_2 measurement is fast and suitable for real-time control systems.

3.4 Equipment for data acquisition and control

The ISE51B ion-selective electrode was used for the measurement of the CO_2 concentration [23]. Electrode potential response to CO_2 concentration is in a semilogarithmic scale, a straight line over two decades of the concentration (5·10–4 g/L to 2·10–2 g/L). For the acquisition of the measured signal of the dissolved CO_2 concentration, the hardware SevenMulti from Mettler Toledo together with LabX direct software was implemented. The control system was realized by means of dSpace 1103 device [22]. Block diagrams of the laboratory measurement and control system are shown in **Figures 1** and **2**.

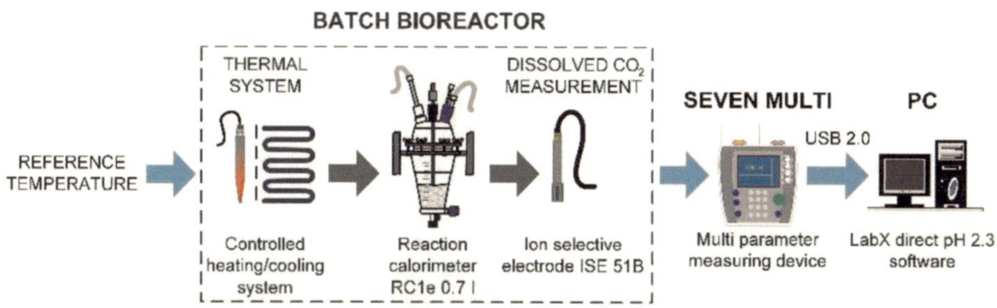

Figure 1.
Laboratory batch bioreactor system for milk fermentation process analysis and modeling [22].

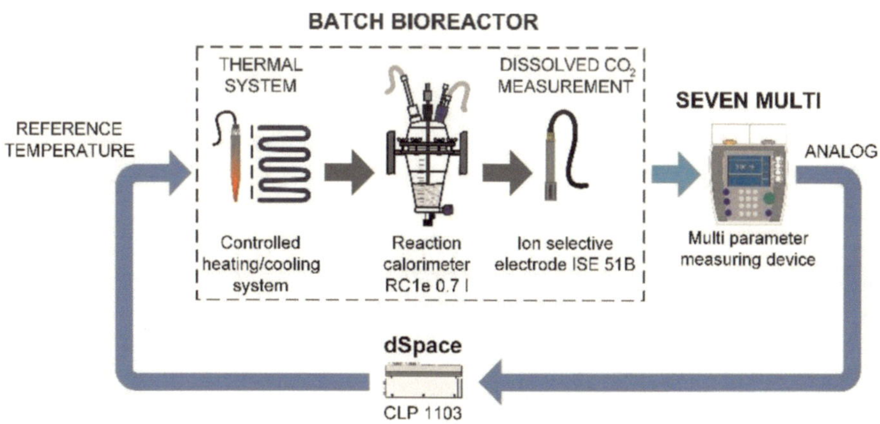

Figure 2.
Laboratory batch bioreactor system for the control of the milk fermentation process [22].

4. Mathematical model of the fermentation process dynamics in a batch bioreactor

The fermentation process is a biological process where the microorganisms cause the source substrate to break down into the final product, that is, the microorganisms consume substrate and, during the growth, produce the product. Microorganisms, substrates, and products are basic substances in all fermentation processes. In the presented study, the fermentation processes in batch bioreactors will be considered.

The fermentation process is a complex process. It can be described by a nonlinear and time-dependent mathematical model with a partially known structure and unknown and time-varying parameters. In the references, we can find different mathematical models of different grades of complexity, which can be used for the description of the fermentation processes in batch bioreactors. Most models are based on the mass balances of the microorganisms, substrate, and fermentation product. A fundamental and most frequently applied mathematical model describes the dynamics of the fermentation process in a batch bioreactor with a system of three nonlinear differential equations. The differential equations are used to calculate the concentration of the substances in the batch bioreactor: microorganisms, substrate, and fermentation product. The differential equations are represented by the Eqs. (1)–(3):

$$\dot{x}_1(t) = \frac{\mu_m\left[1 - \frac{1}{P_i}x_3(t)\right]x_2(t)}{S_m + x_2(t) + \frac{1}{S_i}[x_2(t)]^2}x_1(t) \tag{1}$$

$$\dot{x}_2(t) = -\frac{\mu_m\left[1 - \frac{1}{P_i}x_3(t)\right]x_2(t)}{S_m + x_2(t) + \frac{1}{S_i}[x_2(t)]^2}x_2(t) \tag{2}$$

$$\dot{x}_3(t) = \left[\alpha\frac{\mu_m\left[1 - \frac{1}{P_i}x_3(t)\right]x_2(t)}{S_m + x_2(t) + \frac{1}{S_i}[x_2(t)]^2} + \beta\right]x_3(t) \tag{3}$$

where the following symbols for the concentration of the fermentation process substances are used:

$x_1(t)$: microorganisms' concentration (g/L),

$x_2(t)$: substrate's concentration (g/L),

$x_3(t)$: fermentation product's concentration (g/L),

and the following symbols for the parameters of the mathematical model are introduced:

μ_m: maximum growth rate of the microorganisms (h^{-1}),

P_i: inhibition constant of the fermentation product (g/L),

$S_{m:}$ saturation constant of the fermentation substrate (g/L),

S_i: inhibition constant of the fermentation substrate (g/L),

A: correlation parameter among yield of fermentation product and growth of microorganisms and.

β: correlation parameter that denotes the yield of fermentation product independent of the growth of microorganisms (h^{-1}).

The presented mathematical model can be used for the simulations of the fermentation processes in the batch bioreactors in a case when the temperature stays constant during the execution of the fermentation processes. In the case of the changeable temperature, the presented mathematical model does not allow the simulations of the fermentation processes. In this case, the enhanced mathematical model must be used. The enhanced mathematical model, which considers the influence of the bioreactor's temperature changes on the dynamics of the fermentation process, is described by a system of four differential Eqs. (4)–(7) [20]:

$$\dot{x}_1(t) = \frac{\mu_{m\vartheta}(t)\left[1 - \frac{1}{P_{i\vartheta}(t)}x_3(t)\right]x_2(t)}{S_m + x_2(t) + \frac{1}{S_i}[x_2(t)]^2}x_1(t) \tag{4}$$

$$\dot{x}_2(t) = -\frac{\mu_{m\vartheta}(t)\left[1 - \frac{1}{P_{i\vartheta}(t)}x_3(t)\right]x_2(t)}{S_m + x_2(t) + \frac{1}{S_i}[x_2(t)]^2}x_2(t) \tag{5}$$

$$\dot{x}_3(t) = \left[\alpha\frac{\mu_{m\vartheta}(t)\left[1 - \frac{1}{P_{i\vartheta}(t)}x_3(t)\right]x_2(t)}{S_m + x_2(t) + \frac{1}{S_i}[x_2(t)]^2} + \beta\right]x_3(t) \tag{6}$$

$$\dot{x}_4(t) = \frac{1}{T_{\vartheta cs}}[x_4(t) + u(t)] \tag{7}$$

where added to the symbols in Eqs. (1)–(3):

$x_4(t)$ stands for the temperature of the contents of the bioreactor (°K),

$u(t)$ denotes the reference temperature of the bioreactor's temperature control system (°K),

$T_{\vartheta cs}$ marks the time constant of the 1st order lag term, which represents the mathematical model of the controlled heating system (h) and.

$\mu_{m\vartheta}(t)$, $P_{i\vartheta}(t)$ denote functions that represent the impact of the temperature of the contents of the bioreactor on the parameters of the mathematical model of the fermentation process.

Contrary to the constant parameters μ_m and P_i in Eqs. (1)–(3) are $\mu_{m\vartheta}(t)$, $P_{i\vartheta}(t)$ functions, which depend on the bioreactor's temperature. Function $\mu_{m\vartheta}(t)$ denotes the impact of temperature on the maximum microorganisms' growth rate (h^{-1}), and function $P_{i\vartheta}(t)$ denotes the impact of temperature on the product inhibition constant (g/L). Both functions are identified in such a way that they enable good fitting of the mathematical model's responses to the measured trajectories [20]. Different analytical expressions were tested. After their comparison, the following simple functions were chosen, which can assure proper fitting:

$$\mu_{m\vartheta}(t) = \mu_m\left[1 + k_{\mu m}[x_4(t) - \vartheta_0]\right] \tag{8}$$

$$P_{i\vartheta}(t) = P_i[1 + k_{Pi}[x_4(t) - \vartheta_0]] \tag{9}$$

where.

ϑ_0 is the bioreactor's initial temperature (°K),

$k_{\mu m}$ is the parameter that describes the correlation between the temperature variation and the maximum microorganisms' growth μ_m (°K^{-1}), and.

k_{Pi} is the parameter that describes the correlation between the temperature variation and the product inhibition constant P_i (°K^{-1}).

5. Results of the mathematical model

The parameters μ_m, P_i, S_m, S_i, and α of the mathematical model of the fermentation process depending on the used substances and the principle of the operation mode of the batch bioreactor. For the considered laboratory experimental system, the parameters of the mathematical model were obtained from the measured trajectories of the microorganisms, substrate, and product by the particle swarm optimization. The integral absolute error between measured and simulated trajectories represented the applied performance index [20]. The mathematical model parameters, which were obtained from the experiments of the laboratory kefir production and the subsequent optimization, are shown in **Table 1**.

The Eqs. (1)–(3) represent an autonomous mathematical model, that is, the model without input. This is anticipated because the original industrial versions of batch bioreactors do not allow to be influenced during the execution of the fermentation process. Microorganisms and fermentation substrate are inserted into the bioreactor at the beginning of the fermentation process. The time courses of the concentrations of the fermentation process quantities rest only on the used substances. The initial conditions determine the fermentation process dynamics.

The simulation results of the identified mathematical model of the kefir fermentation process in the laboratory batch bioreactor are shown in **Figure 3**. The trajectories show the response of the concentrations of the microorganisms, fermentation substrate, and fermentation product during the fermentation process, which follows the initial states $x_1(0) = 2.6$ g/L, $x_2(0) = 9.0$ g/L and $x_3(0) = 0.1$ g/L. As expected during the fermentation process, the variables that denote microorganisms and product concentration increased, and the substrate concentration variable decreased. The duration of the modeled fermentation process is approx. 10 hours.

For the simulation of the impact of the changing temperature on the trajectories of the quantities of the fermentation process, the model with the Eqs. 4–9 was used. For the studied fermentation process with data in **Table 1**, the following values of the additional coefficients $k_{\mu m}$ and k_{Pi} were estimated to obtain a good fitting of the measured and simulated trajectories:

$$k_{\mu m} = 0.1°\text{K}^{-1} \quad k_{Pi} = 0.04°\text{K}^{-1} \tag{10}$$

Parameter	Value
Maximum growth rate of the microorganisms	$\mu_m = 0.5\ \text{h}^{-1}$
Inhibition constant of the fermentation product	$P_i = 7.0$ g/L
Saturation constant of the fermentation substrate	$S_m = 0.42$ g/L
Inhibition constant of the fermentation substrate	$S_i = 62.15$ g/L
Correlation parameter among yield of Fermentation product and growth of microorganisms	$\alpha = 0.9\ \frac{g/L}{g/L}$
Correlation parameter that denotes the yield of Fermentation product independent of the growth of microorganisms	$\beta = 0.001\ \text{h}^{-1}$

Table 1.
Identified parameters of the mathematical model of the kefir fermentation process in the laboratory bioreactor.

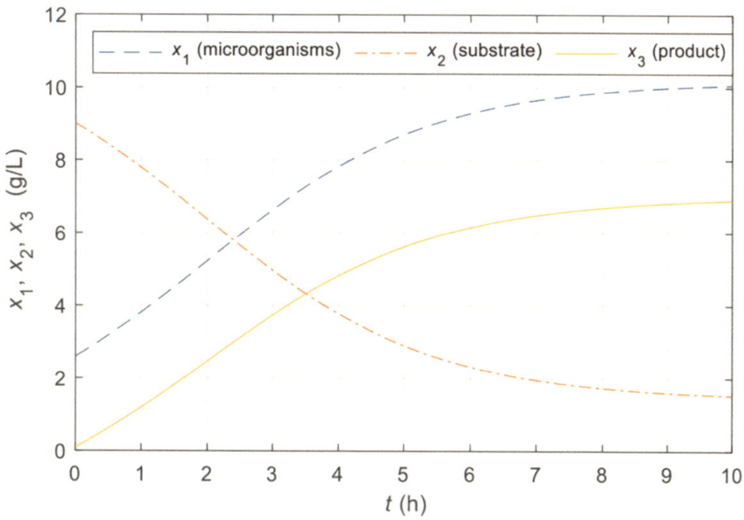

Figure 3.
*Simulation results of the milk fermentation process with kefir grains for constant reference temperature. The parameters of the mathematical model are shown in **Table 1**, presented are concentrations of microorganisms, substrate, and product.*

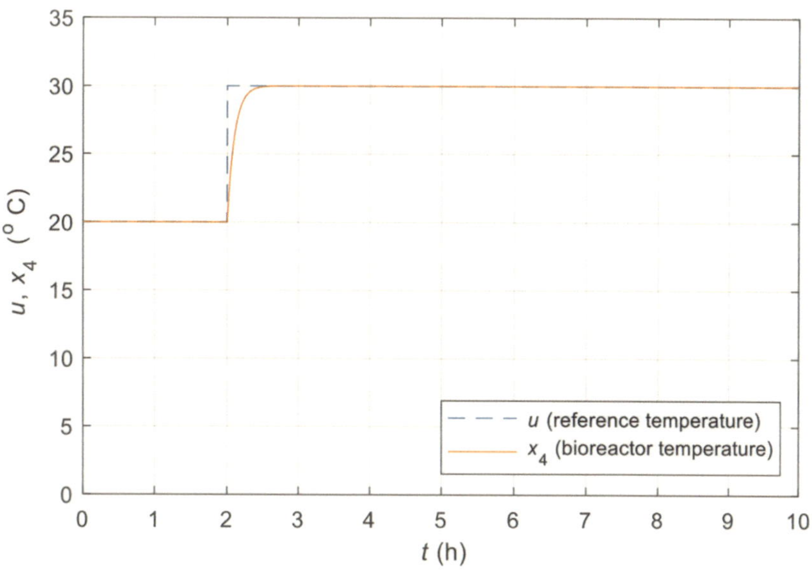

Figure 4.
Simulation results of the milk fermentation process with kefir grains where the step change of the reference temperature at time t = 2 *h is carried out, the parameters of the mathematical model are shown in **Table 1** and Eq. 10, presented are time responses of the reference temperature of the heating control system, and the actual temperature of the contents of the bioreactor.*

The time constant of the controlled heating system was $T_{\vartheta cs}$ = 0.1 h, and the initial bioreactor's temperature was ϑ_0 = 20°C. The time responses of the reference temperature of the heating control system and the actual temperature of the contents of the bioreactor are seen in **Figure 4**. The step change of the reference temperature at time t = 2 h was generated to produce changes in the time response of the mathematical

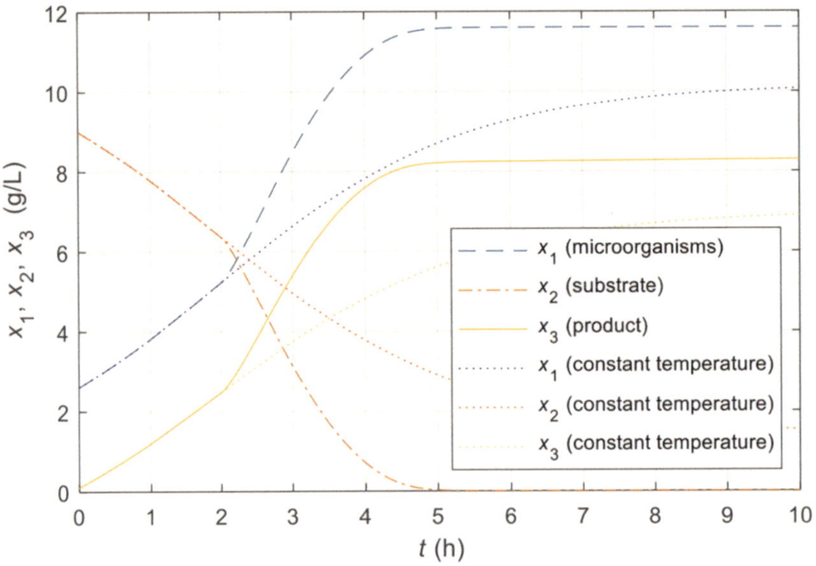

Figure 5.
Simulation results of the milk fermentation process with kefir grains where the step change of the reference temperature at time t = 2 *h is carried out, the parameters of the mathematical model are shown in* ***Table 1*** *and Eq. 10, presented are time responses of concentrations of microorganisms, substrate, and product, dotted/dashed lines show the responses without temperature change.*

model of the fermentation process (4)-(9). The time responses that show the growth of the microorganisms, the consumption of the substrate, and the rising of the fermentation product are shown in **Figure 5**. It is evident that temperature changes generate substantial variations in the dynamics and also in the steady state of the fermentation process. The dotted lines in **Figure 5** represent the time responses of the microorganisms, substrate, and product when no temperature change is made (the same trajectories as in **Figure 1**).

Based on simulations and laboratory tests [18], it was confirmed that changing the reference temperature of the media in the bioreactors significantly impacts the transients and steady state of the fermentation process. This finding confirmed the idea that the bioreactor's heating system could be used to control the fermentation process.

6. Control systems for fermentation process in batch bioreactors

6.1 Conventional control system with a linear controller

The batch bioreactor, which is equipped with a heating system with the possibility of changing the temperature of the medium and with a measuring system for measuring the concentration of the dissolved CO_2, enables the implementation of a conventional feedback control structure. Although the controlled plant is nonlinear and the analysis of its linearized model shows that the variations of parameters of the linearized model are considerable, it was founded that it is possible to use a linear controller with fixed parameters. In this case, the controller must be selected to ensure the stability of the control system in the entire range. The controller chosen in this way will ensure the stability of the control system in the whole operation range. By selecting the performance index and using the optimization method, we can ensure that the controller will provide optimum performance in the broadest possible operating range.

The main disadvantage of this approach is the need for knowledge of the most accurate mathematical model of the fermentation process of the batch bioreactor. Development of an accurate enhanced mathematical model is very complex and requires a lot of time. First of all, the fermentation process at a fixed (unchangeable) temperature must be carried out. The trajectories of the concentrations of microorganisms, substrate, and the product must be measured and nonlinear model parameters (μ_m, P_i, S_m, S_i, α, β) must be calculated. Then, the fermentation with the same initial concentrations must be rerun, this time with the difference that the temperature of the bioreactor mixture is not constant all the time, but that a step change of temperature occurs during fermentation.

6.2 Adaptive control system

6.2.1 Survey of adaptive control systems

The main difficulties in the use of conventional control systems are incomprehension and uncertainty of the mathematical model of the batch bioreactor and, in the case when using a linear controller, changing the parameters of the linearized model of the fermentation process during the operation. The theory of adaptive systems serves as an ideal foundation for developing a control system for batch bioreactors. Adaptive control systems are applicable in case of unknown and changing structures and parameters of the controlled plants and consequently their mathematical models. Adaptive controllers are capable to adapt parameters to the controlled plant variations.

In publications in the field of Systems Theory, it is possible to trace an enormous number of different adaptive and semi-adaptive theories useful for the development of adaptive control systems. Different adaptive approaches are suitable for different controlled plants and require more or fewer controlled plants' quantities to be measured. Controlled plants may be linear or nonlinear, may have a known or unknown structure, may have unknown constant parameters, or their parameters may change during operation. Some adaptive approaches require measurements of all state variables; others require measurements of only one or a few output quantities.

Very many (almost all) adaptive control approaches are derived from the underlying assumption that a controlled plant is linear and has unknown and constant parameters. For such a (simple) case, theoretical proof could be given of the global stability of the entire adaptive system. In practice, however, many adaptive control approaches have shown to be useful, even in cases where the controlled plants are nonlinear, and their parameters are changeable. In this case, the adaptation mechanism adjusts the parameters of the adaptive controller or explicitly generates a control input according to the linearized model of the controlled plant at the current operating point.

Two fundamental concepts are applied to the design of adaptive control systems.

The first approach is based on the separate utilization of three modules: An identification module that identifies the mathematical model of the controlled plant; a setting module that calculates the parameters of the controller on the basis of the identified mathematical model, and a controller module that calculates the corresponding control signal by means of the calculated parameters. This approach is called indirect adaptive control, sometimes also self-tuning control (STC), because initially, it was used primarily for self-tuning of the controller for unknown controlled plants. The benefit of the indirect adaptive control is its modular concept, which enables a connection of various identification techniques (Maximum Likelihood, Least Squares, ...), and different techniques for synthesis of the controller (stochastic, deterministic). The combination of the recursive Least Squares identification method and the pole-shifting controller proved to be one of the most successful. Due to such implementation, the indirect control system is very transparent. During its operation, it is possible to monitor the intermediate results of the adaptive process. The main disadvantage of the indirect adaptive control system, however, is that it is challenging to prove (and ensure) the global stability of the entire adaptive control system. The global stability of the self-tuning adaptive control systems is proven only in the case of more straightforward, less applicable controller synthesis methods [26].

The second approach is based on the use of two components: A reference model and an adaptation mechanism. The reference model prescribes the desired static and dynamic behavior of the controlled plant, and the adaptation mechanism generates control input to ensure that the control plant's output (or/and state variables) follow the output (or/and state variables) of the reference model. This adaptive concept is called direct adaptive control because the input of the controlled plant is determined directly. The other name for these control is also Model Reference Adaptive Control (MRAC). The development of the MRAC approaches is based on assuring the stability of a complete adaptive system. This is the reason that MRAC systems have proven global stability. The majority of direct adaptive control systems can be ranged in one of the following adaptive concepts [27]:

- Model reference adaptive control systems where all controlled plant state space variables must be measured (MRAC with full state access, abbreviated MRAC-FSA) [28],

- Model reference adaptive control systems where only the controlled plant output is measured, and an additional observer is used to observe state space variables (MRAC with an adaptive observer, abbreviated MRAC-AO) [29], and

- Model reference adaptive control systems where only the controlled plant output is measured but no additional observer is needed. In this case, to enable proof of global stability, it is necessary that the controlled plant is "almost" strictly positive real. Therefore, the abbreviation MRAC-ASPR is used [30].

For the needs of the implementation of the batch bioreactor control system, the MRAC-FSA algorithms have proved to be less adequate. Difficulties in applying the MRAC-FSA algorithms are found primarily in the unmeasurability of the necessary state-space variables, which results in the unfulfillment of Erzberger's perfect model following conditions that are required for the implementation of MRAC-FSA control [28].

MRAC-AO algorithms do not require access to all state variables of the controlled plant; the measurement of the output quantity is sufficient. However, when using MRAC-AO systems, accurate knowledge is required of the structure of the mathematical model of the controlled plant. It is necessary to know exactly the maximum order of the mathematical model of the controlled plant (together with the unmodeled dynamics); it is also essential to know the exact relative order of the controlled plant's mathematical model. A limitation is also the requirement that the controlled plant must be a non-minimum phase. All these requirements restrain the use of MRAC-AO algorithms for the control of batch bioreactors.

MRAC-ASPR is the newest of these adaptive concepts. It does not require the usage of an adaptive observer or measurement of all controlled plant variables. MRAC-ASPR is simple for implementation and realization due to time undemanding algorithms. Because of many benefits, the MRAC-ASPR concept was applied to batch bioreactor control.

6.2.2 Theory of MRAC-ASPR system

The presented nonlinear model of the fermentation process can be linearized around the trajectory. This makes it possible to implement the MRAC-ASPR theory, which was originally used for the linear systems described by Eqs. (11) and (12):

$$\dot{\boldsymbol{x}}(t) = \boldsymbol{A}(t)\boldsymbol{x}(t) + \boldsymbol{b}(t)u(t) \tag{11}$$

$$y(t) = \boldsymbol{c}^T(t)\boldsymbol{x}(t) \tag{12}$$

where the symbols are:

$\boldsymbol{x}(t), u(t), y(t)$: controlled plant state-space vector, input, output, and.

$\boldsymbol{A}(t), \boldsymbol{b}(t), \boldsymbol{c}^{\mathrm{T}}(t)$: controlled plant system matrix, input vector, and output vector.

The elements of the system matrix $\boldsymbol{A}(t)$, the input vector $\boldsymbol{b}(t)$, and the output vector $\boldsymbol{c}^{\mathrm{T}}(t)$ are unknown and changeable. It is assumed that:

- The range of the elements of the controlled plant matrices $\boldsymbol{A}(t)$ and vectors $\boldsymbol{b}(t)$ and $\boldsymbol{c}^{\mathrm{T}}(t)$ is bounded,
- All possible pairs $\boldsymbol{A}(t)$ and $\boldsymbol{b}(t)$ are controllable and output stabilizable,
- All possible pairs $\boldsymbol{A}(t)$ and $\boldsymbol{c}^{\mathrm{T}}(t)$ are observable.

The desired dynamics of the batch bioreactor can be described by Eqs. (13) and (14):

$$\dot{\boldsymbol{x}}_{\mathrm{m}}(t) = \boldsymbol{A}_{\mathrm{m}}\boldsymbol{x}_{\mathrm{m}}(t) + \boldsymbol{b}_{\mathrm{m}}u_{\mathrm{m}}(t) \tag{13}$$

$$y_{\mathrm{m}}(t) = \boldsymbol{c}_{\mathrm{m}}^{\mathrm{T}}\boldsymbol{x}_{\mathrm{m}}(t) \tag{14}$$

where the symbols are:

$\boldsymbol{x}_{\mathrm{m}}(t), u_{\mathrm{m}}(t), y_{\mathrm{m}}(t)$: reference model state-space vector, input, output, and.

$\boldsymbol{A}_{\mathrm{m}}(t), \boldsymbol{b}_{\mathrm{m}}(t), \boldsymbol{c}_{\mathrm{m}}{}^{\mathrm{T}}(t)$: reference model system matrix, input vector, and output vector.

The reference model determines the desired input-output relation. Therefore, it is not necessary for the reference model order to be the same that the controlled plant order—it could be smaller as written in Eq. (15):

$$\dim[x_{\mathrm{m}}(t)] \ll \dim[x(t)] \tag{15}$$

The deviation between the reference model and controlled plant dynamics is described by the error between both outputs:

$$e_y(t) = y_m(t) - y(t) \tag{16}$$

The Lyapunov stability theory is used for the derivation of the adaptive control algorithm (17) [31, 32]. This algorithm calculates such input of the controlled plant $u(t)$ that assures that the output of the controlled plant $y(t)$ will follow the reference model output $y_m(t)$ in case of unknown controlled plant model $\boldsymbol{A}(t)$, $\boldsymbol{b}(t)$, and $\boldsymbol{c}^T(t)$.

$$u_p(t) = K_e(t)e_y(t) + \boldsymbol{K}_x(t)\boldsymbol{x}_m(t) + K_u(t)u_m(t) \tag{17}$$

where the additional symbols are:
$\boldsymbol{K}_x(t)$, $K_u(t)$: control gains, and.
$K_e(t)$: output feedback parameter.
Variables $e_y(t)$, $\boldsymbol{x}_m(t)$, and $u_m(t)$ and parameters $K_e(t)$, $\boldsymbol{K}_x(t)$, and $K_u(t)$ can be rewritten with control variables vector $\boldsymbol{r}(t)$ and adaptive gains vector $\boldsymbol{K}(t)$:

$$\boldsymbol{K}(t) = [K_e(t) \;\; \boldsymbol{K}_x(t) \;\; K_u(t)] \tag{18}$$

$$\boldsymbol{r}^T(t) = \left[e_y(t) \quad \boldsymbol{x}_m(t) \quad u_m(t)\right] \tag{19}$$

The adaptive gains vector $\boldsymbol{K}(t)$ can be calculated with the proportional-integral algorithm:

$$\boldsymbol{K}(t) = \boldsymbol{K}_p(t) + \boldsymbol{K}_i(t) \tag{20}$$

where symbols denote:
$\boldsymbol{K}_p(t)$: proportional term $\boldsymbol{K}_p(t)$, and.
$\boldsymbol{K}_i(t)$: integral term.
The integral term $\boldsymbol{K}_i(t)$ guarantees convergence and the proportional term $\boldsymbol{K}_p(t)$ generates a fast response. The MRAC-ASPR algorithm increases the values of the adaptive gains vector $\boldsymbol{K}(t)$ if the error increase and decreases if the error decrease [28]. Both terms of the adaptive gains vector $\boldsymbol{K}(t)$ can be calculated by Eqs. (21) and (22) [28]:

$$\boldsymbol{K}_p(t) = e_y(t)\boldsymbol{r}^T(t)\,\boldsymbol{T}' \tag{21}$$

$$\dot{\boldsymbol{K}}_i(t) = e_y(t)\boldsymbol{r}^T(t)\,\boldsymbol{T} \tag{22}$$

where symbols denote:
$\boldsymbol{T}$': positive semi-definite matrix, and $\boldsymbol{T}$: positive definite matrix.
Considering (22) for the calculation of $\boldsymbol{K}_i(t)$, it is obvious that the proposed integral term in the presence of disturbances, whenever perfect following is not possible, may reach unnecessarily large values, or may even diverge. In order to improve the convergence of the adaptive system, the following modification of the integral term was proposed [30]:

$$\dot{\boldsymbol{K}}_i(t) = e_y(t)\boldsymbol{r}^T(t)\,\boldsymbol{T} - \sigma\boldsymbol{K}_i(t) \tag{23}$$

where the σ-term is used to evade integral gains divergence in the presence of a disturbance.

The origin theory for the development of the MRAC-ASPR control systems is described in detail in [30].

7. Results of the control systems

Both considered control strategies were tested for the control of the fermentation process in the batch bioreactor with the data in **Table 1**. The treated batch bioreactor was equipped with a final control system for heating/cooling the bioreactor's contents and with a measurement system that measures the concentration of the dissolved CO_2 in the bioreactor's vessel. The input of the thus-formed controlled plant represents the reference signal for the heating/cooling system, and the output of the controlled plant is the signal that corresponds to the dissolved CO_2 concentration.

7.1 Reference trajectory of the dissolved CO_2

The quantity and quality of the product in the batch bioreactor are decisively dependent on the trajectories of the biological quantities in the fermentation process, affecting the kinetics of the bioprocess [1, 3, 12, 22]. In today's used industrial batch bioreactors without a closed-loop control system, the course of the fermentation process depends only on the substances that were placed in the bioreactor before the beginning of the fermentation process and cannot be influenced during the fermentation process. The developed and presented control system makes the changing of the time responses of the biological quantities possible.

With this new possibility, the presented control system will enable the fermentation process to be carried out, where the time courses of concentrations of biomass, substrate and fermentation result will be the same as the prescribed reference trajectories. It makes sense that the reference trajectories are chosen in such a way that the fermentation process gives as much as possible and as much quality product as possible in the shortest possible time. There are many publications that describe in more detail how to select reference trajectories [12]. In the presented contribution, we did not theoretically deal with the problem of choosing reference trajectories but limited ourselves only to choosing the reference trajectory of the concentration of dissolved CO_2. The reference trajectory was chosen based on the experience of operators operating batch bioreactors.

The basis for the selection of the reference signal is the time course of the corresponding quantity of the unregulated fermentation process. When choosing a reference trajectory, we must take into account the biological limitations of the fermentation process. The dynamics of the reference signal must be such that the regulation system of the bioreactor can realize it. For the tested fermentation process in the available bioreactor, the following reference trajectory $r(t)$ for the dissolved CO_2 was chosen:

$$r(t) = 0.8\left(1 - e^{-t/1.5}\right) \tag{24}$$

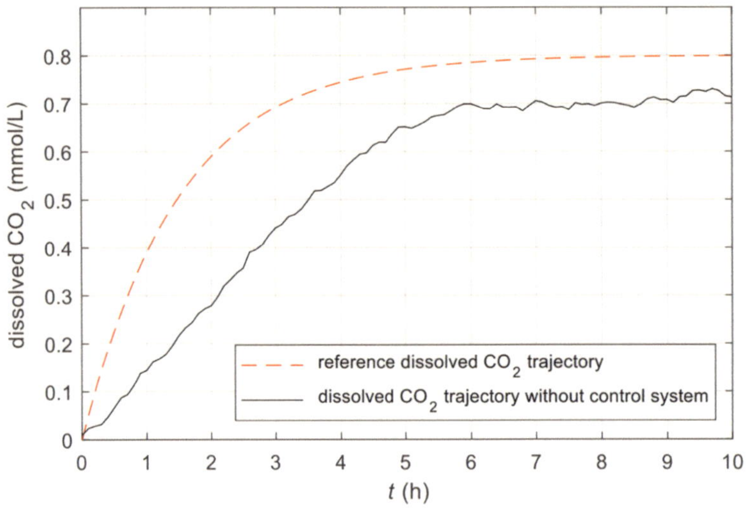

Figure 6.
Actual and the reference trajectory of the CO_2 concentration; data of the bioreactor are in ***Table 1***.

The time response of the CO_2 concentration of the unregulated bioreactor and the chosen suitable reference trajectory for the control system is shown in **Figure 6**.

7.2 Conventional control system with a linear controller

The use of linear controllers makes sense in the case of a batch bioreactor since the analysis of a linearized mathematical model in the vicinity of the trajectory showed a relatively small range of variations in the model's parameters. In such cases, the controller must be set to ensure stability and good transients of the control system in the most problematic operating range. Of course, the main limitation in the use of such controllers is the requirement to identify the mathematical models of the controlled plants.

A simple PI controller with transfer function $G_{\text{PI}}(s)$ [33] is used to demonstrate the usefulness and efficiency of the conventional control system with a linear controller for the batch bioreactor control.

$$G_{\text{PI}}(s) = k_{\text{p}} \frac{sT_{\text{i}} + 1}{sT_{\text{i}}} \tag{25}$$

where k_{p} is the gain and T_{i} is the time constant of the PI controller described with the transfer function $G_{\text{PI}}(s)$.

The controller design is based on the known nonlinear and linearized mathematical models. The design procedure was divided into two phases:

- First, the time constant T_{i} of an integral part of the PI controller was determined regarding the unstable region of the operating range. The selected time constant was determined on the basis of the linearized mathematical model, and assures a stable operation for small deviations around the entire trajectory.

- Second, the gain k_p of the proportional part of the PI controller was calculated using the differential evolution. Integral time square cost function J of the output error variable and input variable (26) was used to determine the optimal value of the controller's gain [21].

$$J = \int_0^{t_f} \left\{ Q\,[y^*(t) - y(t)]^2 + R[u^*(t) - u(t)]^2 \right\} dt \tag{26}$$

where.

$u(t)$ is the input variable of the controlled plant (i.e., the reference temperature of the bioreactor's temperature control system (°K)),

$y(t)$ is the output variable of the controlled plant (i.e., the output of the measurement system for the dissolved CO_2 concentration (mmol/L)) mathematical model,

$u^*(t), y^*(t)$ are the nominal input and output variables of the controlled plant,

Q, R are the weighting parameters of the quadratic cost function, and.

t_f - final time.

In the optimization calculations, it is assumed that the nominal input variable is constant during the entire fermentation process and is equal to the temperature of the bioreactor's filling at the beginning of the fermentation process ϑ_0 (°K). For the chosen cost function's parameters Q, R, t_f, and the constant temperature of the bioreactor's filling ϑ_0:

$$Q = 1\; R = 0.1\; t_f = 10\,h\;\; \vartheta_0 = 292°K \tag{27}$$

the following parameters of the PI controller were calculated:

$$k_p = 22.0\; T_{PI} = 1.5\,\text{h} \tag{28}$$

The corresponding block diagram of the control system is shown in **Figure 7**. The obtained results are presented in **Figures 8** and **9**. The time response of the actual

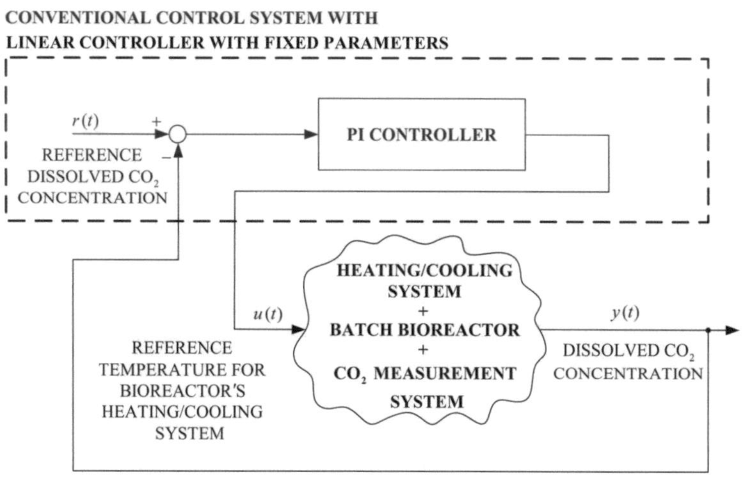

Figure 7.
Block diagram of the conventional control system with linear controller.

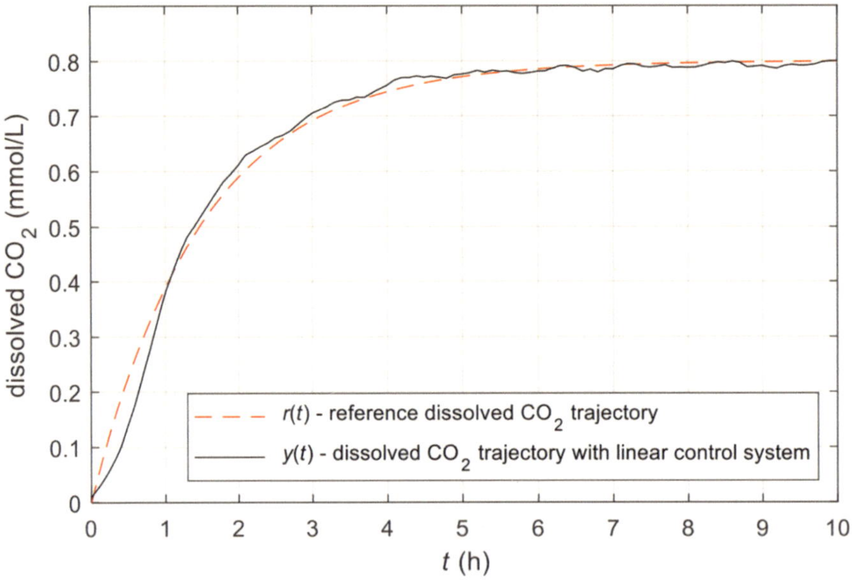

Figure 8.
*Time response of the actual and reference-dissolved CO_2 concentration of a batch bioreactor with the conventional control system with a linear controller with calculated parameters; data of the bioreactor are in **Table 1**, controller parameters in ref. (28).*

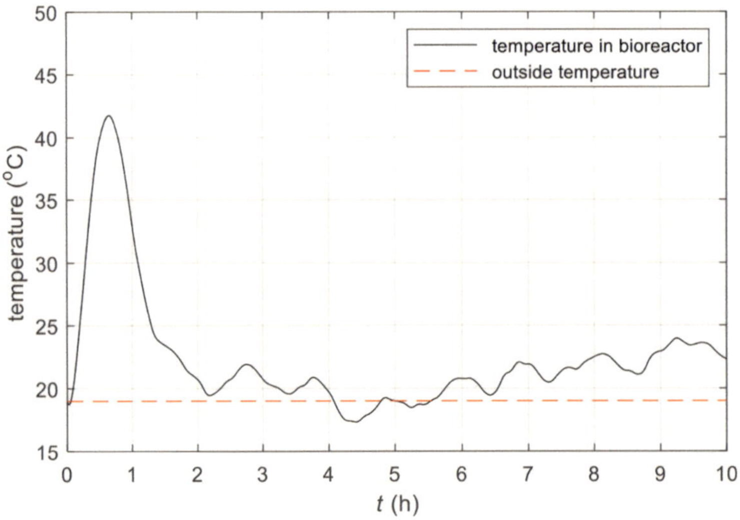

Figure 9.
*Time response of the batch bioreactor's inner temperature and the constant outside temperature when the conventional control system with a linear controller with calculated parameters was used; data of the bioreactor are in **Table 1**, controller parameters in ref. (28).*

dissolved CO_2 concentration of a batch bioreactor controlled with a conventional PI controller, together with the reference trajectory, is shown in **Figure 8**. The bioreactor's inner temperature $x_4(t)$, as a consequence of the control of the heating/cooling system and the outside temperature ϑ_0 are presented in **Figure 9**.

7.3 Adaptive control system

A simple adaptive control system based on the MRAC-ASPR control theory was used for the batch bioreactor's control implementation. The block diagram of the adaptive control system for the batch bioreactor's control is shown in **Figure 10**.

The proposed adaptive control system represents a kind of model reference adaptive control system—the adaptation mechanism and control law will ensure that even in the case of unknown and changeable parameters of the controlled plant, the course of the fermentation process will be as close as possible with the dynamics of the selected reference model. The reference model was determined based on the selected reference signal. For the selected reference signal (24), we chose 1st order lag with gain k_{rm} = 0.8 and the time constant T_{rm} = 1.5 h as the reference model.

The speed of the adaptation mechanism depends on the matrix parameters ***T'*** and ***T***. To ensure the stability of the adaptive system, it is required that the matrix ***T'*** is

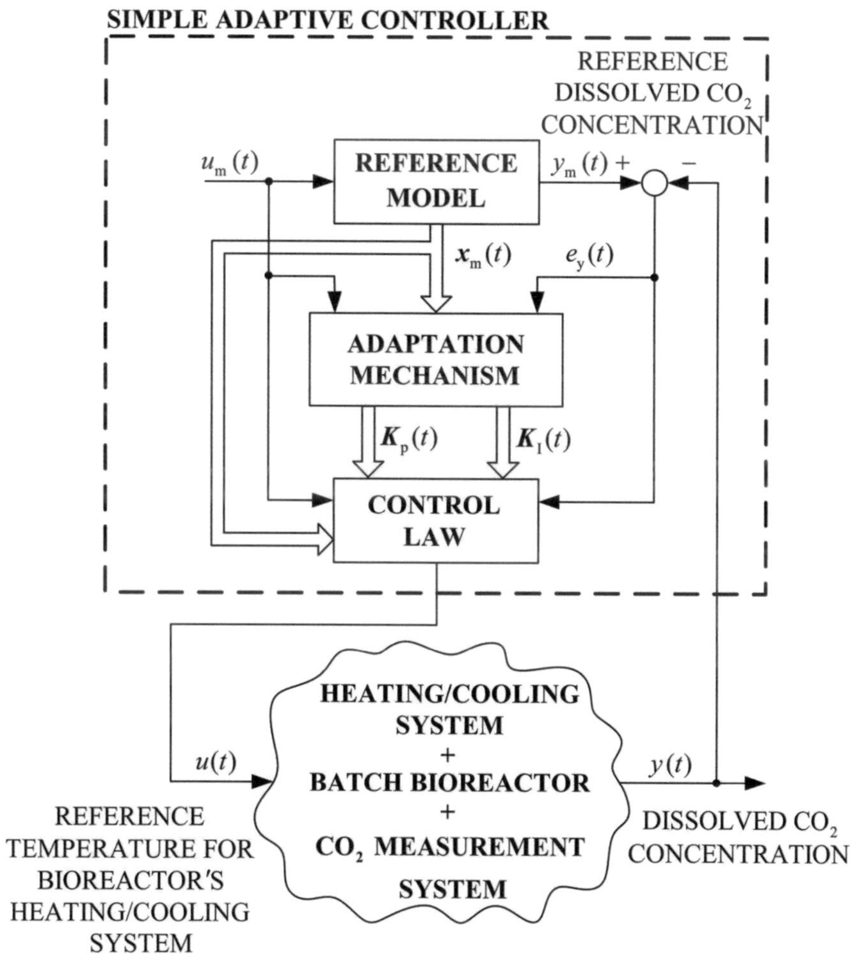

Figure 10.
Block diagram of the adaptive control system.

positive semi-definite and the matrix $\boldsymbol{T}$ is a positive definite matrix. Numerical simulations were used to choose the parameters:

$$\boldsymbol{T} = 4 \cdot 10^3 \begin{bmatrix} 1 & 0 & 0 \\ 0 & 1 & 0 \\ 0 & 0 & 1 \end{bmatrix} \boldsymbol{T}' = 4 \cdot 10^3 \begin{bmatrix} 1 & 0 & 0 \\ 0 & 1 & 0 \\ 0 & 0 & 1 \end{bmatrix} \tag{29}$$

To avoid divergence of an integral part in the presence of disturbance the following σ-term was used:

$$\sigma = 0.95 \tag{30}$$

The results of the simple adaptive control technique for the fermentation process in the batch bioreactor are shown in **Figures 11** and **12**. **Figure 11** shows the reference and the actual time response of the dissolved CO_2 concentration when a simple adaptive controller was used. The obtained results show that with the proposed adaptive regulation system, we can achieve the time course of the dissolved CO_2 concentration will be very similar (almost the same) as the prescribed reference time course. The regulation system ensures the same course of the fermentation process even in the case of small variations in the initial values of the quantity and quality of the fermentation components (different charges).

The control system ensures the same dynamics of the fermentation process by changing the temperature in the bioreactor. **Figure 12** shows the required temperature in the bioreactor. It can be seen from **Figure 12** that the temperature required for the control remains within an acceptable temperature range.

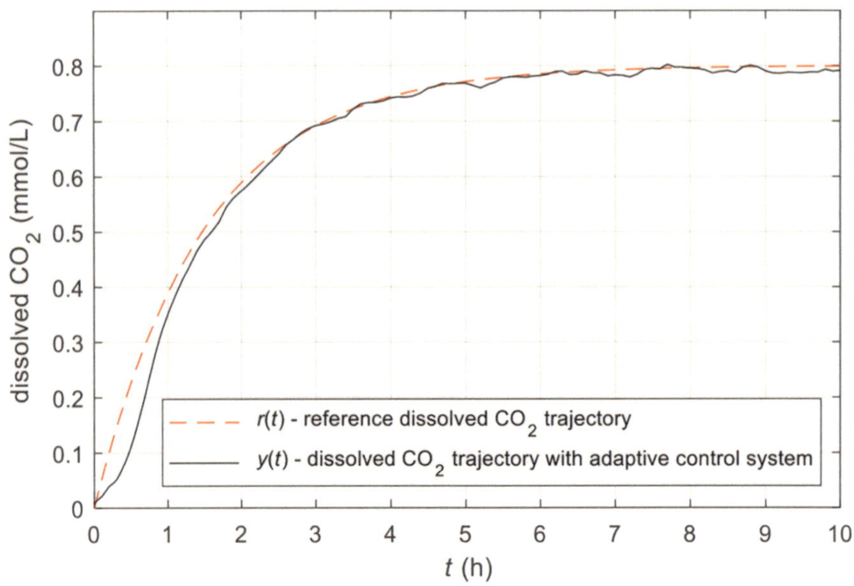

Figure 11.
Time response of the actual and reference dissolved CO_2 concentration of a batch bioreactor with the simple adaptive control system; the data of the bioreactor are shown in ***Table 1****, and adaptation mechanism parameters in [29, 30].*

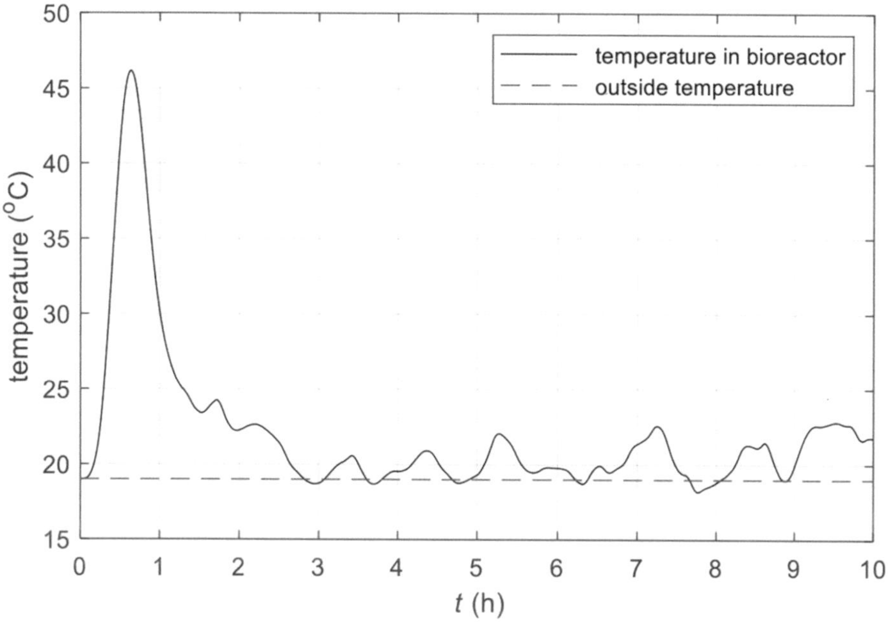

Figure 12.
*Time response of the batch bioreactor's inner temperature and the constant outside temperature when the simple adaptive control was used; the data of bioreactor are shown in **Table 1**, and adaptation mechanism parameters in [29, 30].*

8. Discussion

At first sight, the results obtained with both presented control systems are excellent and very similar. Almost identical dynamics of the fermentation process were obtained, as defined by the reference trajectory. To achieve these responses, acceptable changes were requested in the bioreactor's inner temperature.

To obtain better insight into the performances of the control systems, the evaluation based on the performance index is meaningful. The integral quadratic performance index, the same as the cost function shown in Ref. (26), was used for the comparison. The same parameters of the performance index as for the PI-controller optimization were used to estimate control quality. The calculated performance indexes for the conventional control system with PI controller and the simple adaptive control system are presented in **Table 2**.

It is surprising that the performance indexes of conventional PI control system and advanced adaptive control system are very similar. From the point of view of the quality of the control systems, there is no significant difference between the PI control system and the adaptive control system. However, there is a significant difference in

Control Concept	**Value**
Conventional control system with PI controller	$J = 4.2496$
Simple adaptive control system	$J = 4.2864$

Table 2.
Performance indexes for the studied Bioreactor's control systems.

the use of both control systems. Using the PI control system is very time-consuming for bioreactor operators. To determine the optimal parameters of the PI controller, a precise determination of the mathematical model of the fermentation process is necessary. This is very demanding and time-consuming. The adaptive controller has a great advantage, as it automatically adapts to the controlled system. Operators do not need to set the parameters of the controller, which significantly simplifies and speeds up their work.

Despite the slightly worse performance index, the proposed adaptive control approach presents a much better choice for the development of the practical control system for the batch bioreactor.

9. Conclusion

There are two important contributions to the presented work:

- It has been shown that batch bioreactors, which represent the simplest and cheapest type of bioreactors and which basically allow only autonomous (uncontrolled) operation, can be turned into a closed-loop control system with a little additional equipment. The basic condition is that the bioreactors have a heating/cooling system, through which the controller can influence the course of the fermentation process, and

- It has been shown that the developed adaptive control system represents a very effective control for batch bioreactor operation, even in the case of an unknown, nonlinear, and time-variable mathematical model of the fermentation process.

The chapter shows the possibility of using two different control concepts: conventional PI control system and advanced adaptive control system. The adaptive control gives almost the same results as the PI control system, whose parameters were offline tuned with the optimization, on the basis of the identified mathematical model of the fermentation process.

Adaptive control system proved to be significantly more suitable for the control of batch bioreactors. Its main advantage is very simple to use, which does not require a prior setting of parameters, but the regulator itself adapts to the characteristics of the controlled fermentation process. In such a way, the batch bioreactors, which are easy to fabricate and maintain (and are, therefore, consequently less flexible to control), thanks to the advanced control theory, easily and cheaply acquire the possibility to improve their performance significantly. The shorter fermentation time and higher quality of the obtained products are guaranteed, which is reflected in their greater efficiency of operation.

In the future, we plan to continue the work we have started. The goal is to simplify the realization of the control system and to carry out testing on as many different batch bioreactors as possible.

Author details

Jožef Ritonja[1*], Andreja Goršek[2], Darja Pečar[2], Darius Andriukaitis[3] and Boštjan Polajžer[1]

1 Faculty of Electrical Engineering and Computer Science, University of Maribor, Maribor, Slovenia

2 Faculty of Chemistry and Chemical Engineering, University of Maribor, Maribor, Slovenia

3 Faculty of Electrical and Electronics Engineering, Kaunas University of Technology, Kaunas, Lithuania

*Address all correspondence to: jozef.ritonja@um.si

References

[1] Henson MA. Exploiting cellular biology to manufacture high-value products – Biochemical reactor modelling and control. IEEE Control Systems Magazine. 2006;**26**:54-62

[2] Shuler ML, Kargi F. Bioprocess Engineering: Basic Concepts. 2nd ed. Englewood Cliffs, NJ: Prentice Hall; 2002

[3] Cinar A, Parulekar SJ, Undey C, Birol G. Batch Fermentation - Modelling, Monitoring and Control. New York: Marcel Dekker Inc.; 2003

[4] Blanch HW, Clark DS. Biochemical Engineering. New York: Marcel Dekker, Inc.; 1997

[5] Mazenc F, Jiang Z-P. Global output feedback stabilization of a Chemostat with an arbitrary number of species. IEEE Transactions on Automatic Control. 2010;**55**(11):2570-2575. DOI: 10.1109/TAC.2010.2060246

[6] Yan X, Bergstrom DJ, Chen XB. Modeling of cell cultures in perfusion bioreactors. IEEE Transactions on Biomedical Engineering. 2012;**59**(9): 2568-2575. DOI: 10.1088/0957-0233/22/7/072001

[7] Coutinho D, Vande Wouwer A. A robust non-linear feedback control strategy for a class of bioprocesses. IET Control Theory & Applications. 2013;**7**(6): 829-841

[8] Mairet F, Gouzé J-L. Hybrid control of a bioreactor with quantized measurements. IEEE Transactions on Automatic Control. 2016;**61**(5):1385-1390. DOI: 10.1109/TAC.2015.2471735

[9] de Andrade GA, Pagano DJ, et al. Distributed sliding mode control of pH in tubular Photobioreactors. IEEE Transactions on Control Systems Technology. 2016;**24**(4):1-14. DOI: 10.1109/TCST.2015.2480840

[10] Battista HD, Picó-Marco E, Santos-Navarro FN, Picó J. Output feedback linearization of Turbidostats after time scaling. IEEE Transactions on Control Systems Technology. 2019;**27**(4):1-9. DOI: 10.1109/TCST.2018.2834882

[11] Casenave C, Perez M. "Antiwindup input–output linearization strategy for the control of a multistage continuous fermenter with input constraints". IEEE Transactions on Control Systems Technology. 2019;**28**(3):766-775. DOI: 10.1109/TCST.2019.2892932

[12] Peroni CV, Kaisare NS, Lee JH. Optimal control of a fed-batch bioreactor using simulation-based approximate dynamic programming. IEEE Transactions on Control Systems Technology. 2005;**13**(5):786-790. DOI: 10.1109/TCST.2005.852105

[13] Romoli S, Scaglia GJE, et al. Control of a fed-batch fermenter based on a linear algebra strategy. IEEE Latin America Transactions. 2014;**12**(7):1206-1213. DOI: 10.1109/TLA.2014.6948854

[14] Rodriguez AE, Munoz JAT, et al. Robust control for cultivation of microorganisms in a high density fed-batch bioreactor. IEEE Latin America Transactions. 2015;**13**(6):1927-1933. DOI: 10.1109/TLA.2015.7164219

[15] Estakhrouiyeh MR, Vali M, Gharaveisi A. Application of fractional order iterative learning controller for a type of batch bioreactor. IET Control Theory & Applications. 2016;**10**(12):1374-1383. DOI: 10.1049/iet-cta.2015.1268

[16] Arévalo H, Snáchez F, et al. Gain-scheduled oxygen concentration control system for a bioreactor. IEEE Latin

America Transactions. 2018;**16**(11): 2689-2697. DOI: 10.1109/TLA.2018.8795109

[17] Farias G, Fabregas E, et al. Automatic control for the production of alginate by Azotobacter Vinelandii. IEEE. Access. 2019;7:168606-168612. DOI: 10.1109/ACCESS.2019.2954180

[18] Goršek A, Ritonja J, Pečar D. Mathematical model of CO2 release during milk fermentation using natural kefir grains. Journal of the Science of Food and Agriculture. 2018;**98**: 4680-4684. DOI: 10.1002/jsfa.9001

[19] Ritonja J, Gorsek A, Pecar D. Control of Milk fermentation in batch bioreactor. Elektronika Ir Elektrotechnika. 2020; **26**(1):4-9. DOI: 10.5755/j01.eie.26.1.23377

[20] Ritonja J, Goršek A, Pečar D, Petek T, Polajžer B. Dynamic Modeling of the impact of temperature changes on CO2 production during Milk fermentation in batch bioreactors. Food. 1809;**2021**:10. DOI: 10.3390/foods10081809

[21] Ritonja J, Goršek A, Pečar D. Use of a heating system to control the probiotic beverage production in batch bioreactor. Applied Sciences. 2021;**11**(1):84. DOI: 10.3390/app11010084

[22] Ritonja J. Adaptive control of CO2 production during Milk fermentation in a batch bioreactor. Mathematics. 2021; **9**:1712. DOI: 10.3390/math9151712

[23] Ritonja J, Goršek A, Pečar D. Model reference adaptive control for Milk fermentation in batch bioreactors. Applied Sciences. 2020;**10**(24):9118. DOI: 10.3390/app10249118

[24] Goršek A, Tramšek M. Kefir grains production—An approach for volume optimization of two-stage bioreactor system 2008. Biochemical Engineering Journal. 2008;**42**(2):153-158. DOI: 10.1016/j.bej.2008.06.009

[25] Zosel J, Woelßner. Topical review: The measurement of dissolved and gase-ous carbon dioxide concentration. Measurement Science and Technology. 2011;**22**:45. DOI: 10.1088/0957-0233/22/7/072001

[26] Isermann R, Lachmann K-H, Matko D. Adaptive Control Systems. New York: Prentice Hall International; 1992

[27] Ritonja J. Adaptive stabilization for generator excitation system. COMPEL - The International Journal for Computation and Mathematics in Electrical and Electronic Engineering. 2011;**30**(3):1092-1108. DOI: 10.1108/03321641111111022

[28] Landau YD. Adaptive Control: The Model Reference Approach. New York, NY, USA: Marcel Dekker, Inc.; 1979

[29] Narendra KS, Annaswamy AM. Stable Adaptive System. Englewood Cliffs, New Jersey: Prentice Hall Inc.; 1989

[30] Kaufman H, Bar-Khana I, Sobel K. Direct Adaptive Control Algorithms. New York: Springer Verlag; 1993

[31] Parks P. Liapunov redesign of model reference adaptive control systems. IEEE Transactions on Automatic Control. 1966;**11**(3):362-367. DOI: 10.1109/TAC.1966.1098361

[32] LaSalle J. Stability of nonautonomous systems. Nonlinear Analysis Theory Methods and Applications. 1981;**1**(1):83-88. DOI: 10.1016/0362-546X(76)90011-0

[33] Åström KJ, Hagglund T, Controllers PID. Theory, Design, and Tuning. North Carolina: The International Society for Measurement and Control; 1988

Chapter 8

Enhancing Resilience in Cooperative Systems against Cyber-Attacks: A Defense Framework through Adaptive Network Reconfiguration and Digital Twin

Azwirman Gusrialdi, Deepalakshmi Babu Venkateswaran and Zhihua Qu

Abstract

The book chapter presents a solution to enhance the resilience of multi-agent cooperative systems against cyber-attacks. Specifically, the threat involves adversaries aiming to prevent the cooperative systems from achieving consensus by launching false data injection (FDI) or denial-of-service (DoS) attacks on the communication network. First, a distributed algorithm is proposed to identify critical edges, via online estimation/learning of the network's parameters, whose removals disconnect the network. An adaptive network reconfiguration algorithm is then presented to remove those critical edges, thus ensuring the resilient operation of the cooperative systems against arbitrary DoS attacks. Additionally, a digital twin, interconnected with the cooperative systems, is introduced to simultaneously ensure the resiliency of the cooperative systems against FDI attacks and increase the redundancy of the network. Finally, it is illustrated how the proposed defense framework also allows for detecting and identifying in real-time both attack vectors in the network. The efficacy of the proposed solution is demonstrated through numerical simulations, showcasing their effectiveness in protecting the cooperative systems against cyber threats.

Keywords: multi-agent cooperative systems, consensus algorithm, cyber-attacks, network reconfiguration, attack detection and identification

1. Introduction

Cooperative systems are a collection of autonomous agents that interact/cooperate with each other to make their own decision (actions) so that the overall system achieves a common goal and/or the overall system's performance is optimized without the need for a central entity. In practice, the individual agent can be distributed generator in a power system, an autonomous robot in a robotic swarm, or an autonomous vehicle in an intelligent transportation system, while the interaction can take place via wireless communication. The local interaction rule makes the cooperative systems appealing for real-time decision-making in large-scale and critical systems due to their scalability and low latency.

The consensus problem is a canonical issue in cooperative systems in which the agents aim to achieve agreement, via a consensus algorithm, on specific quantities of interest, such as voltage, distance, or velocity [1]. Various coordination problems in diverse domains, including power systems, transportation systems, robotic swarms, and federated learning, can be framed as consensus problems [2–5]. Furthermore, the consensus algorithm also serves as a fundamental component in numerous distributed optimization algorithms [6, 7].

While the introduction of a communication network plays a crucial role in realizing cooperation among the agents, its use comes at the price of making the cooperative system vulnerable to cyber-attacks, which may prevent the systems from achieving consensus and, in the worst case, destabilize the overall systems [8, 9]. Recent history has witnessed catastrophic cyber-attacks, such as the 2015 Ukraine power grid incident, causing a 6-hour blackout [10]. Specifically, the attacker may modify the information being exchanged between the agents. Thus destroying data integrity, e.g., by launching false data injection (FDI attacks), or block the communication channels causing data unavailability, e.g., by launching denial-of-service (DoS) attacks. In addition, the attacker can also passively eavesdrop on the communication channel to acquire confidential information about the agents [11].

In order to unlock the potential economic impact of cooperative systems, it is thus imperative to ensure the resilient operation of the cooperative systems against unknown cyber-attacks. This book chapter focuses on cooperative systems under DoS or FDI attacks. The existing strategies for resilient consensus against DoS attacks impose restrictive assumptions on the frequency and duration of attacks [12, 13], which are hard to guarantee in practice. A strategy to ensure resilient consensus against FDI attacks is by removing a fixed number of extreme local values that an agent received from the other agents [14]. However, this strategy requires high network connectivity and cannot guarantee the agents to converge to the attack-free consensus value even in the absence of attacks. Another potential strategy is by identifying the compromised communication links and subsequently removing them [15]. However, this approach also requires high network connectivity, which depends on the number of compromised links, and more importantly, the system's stability may have been destroyed during the identification process.

This book chapter presents a novel defense framework that eliminates restrictions imposed in the existing strategies discussed previously. First, to defend against DoS attacks, a distributed algorithm is presented to determine the critical edge in the network, which is an edge whose removal disconnects the network. A novel network reconfiguration algorithm is then proposed to remove the existence of the critical edge and thus ensure that the network remains connected under arbitrary DoS attacks. Furthermore, a digital twin is designed and interconnected with the cooperative systems. The digital twin not only makes the cooperative system resilient against FDI attacks but also helps to remove the critical edges in the network. The proposed defense framework also facilitates real-time and distributed detection and identification of DoS or FDI attacks in the network. Finally, a numerical example is provided to illustrate the proposed defense framework.

2. Preliminaries

Let $\mathbb{R}$ be the set of real numbers; vector $\mathbf{1}_n \in \mathbb{R}^n$ denotes the vector of all ones, $\mathbf{0}_n \in \mathbb{R}^n$ represents the vector of all zeros, and $I_n \in \mathbb{R}^{n \times n}$ denotes an $n \times n$ identity matrix. The cardinality of a set $\mathcal{N}$ is denoted by $|\mathcal{N}|$. Superscript $''T''$ represents the transpose of a matrix or a vector.

Let $\mathcal{G} = (\mathcal{V}, \mathcal{E})$ be an undirected graph with a set of nodes $\mathcal{V} = \{1, \cdots, n\}$ and a set of edges $\mathcal{E} \subset \mathcal{V} \times \mathcal{V}$. An edge $(i,j) \in \mathcal{E}$ denotes that node j can receive information from node i. Since the graph is undirected, we have $(i,j) \in \mathcal{E} \Leftrightarrow (j,i) \in \mathcal{E}$. A path from node i to node j is a sequence of consecutive edges $\{(i,m), (m,q), \cdots, (l,j)\}$ with distinct vertices. A path that starts from a given node and ends at the same node is called a cycle. An undirected graph $\mathcal{G}$ is connected if a path exists between every pair of nodes. The distance between two nodes in a graph is the length/number of consecutive edges connecting them. The neighbor set of node i is defined as $\mathcal{N}_i = \{j, (j,i) \in \mathcal{E}, j \neq i\}$.

3. Multi-agent cooperative system

Consider a multi-agent cooperative system (MACS) Σ_s consisting of n agents whose individual dynamics are given by

$$\dot{x}_i = u_i, \quad i = \{1, \cdots, n\} \tag{1}$$

where $x_i \in \mathbb{R}$ is the state of physical variables to be controlled and u_i is the control input. The agents can exchange information with each other via a communication network whose network topology is modeled by a graph $\mathcal{G}$ whose nodes represent the agents. It is assumed that the communication between a pair of agents is bidirectional, that is graph $\mathcal{G}$ is undirected and an edge $(i,j) \in \mathcal{E}$ denotes that agents i and j can exchange information with each other.

In this book chapter, we focus on a consensus problem where the goal is to design a local interaction rule u_i for an individual agent, which depends on its own and its neighboring agents' states, such that the state of all the agents in the network reach a consensus, i.e.,

$$x_i \to c, \forall i \text{ and } c \neq 0. \tag{2}$$

To this end, each agent executes the following consensus law [1].

$$u_i = \sum_{j \in \mathcal{N}_i} (x_j - x_i). \tag{3}$$

Defining a vector $x = [x_1, \cdots, x_n]^T$, the dynamics (1) under consensus law (3) can be written in a compact form as

$$\dot{x} = Ax = -Lx, \tag{4}$$

where L is called a Laplacian matrix associated with the graph $\mathcal{G}$. If the communication network topology $\mathcal{G}$ is connected, then all the agents' states reach a consensus whose value is equal to the average of the initial conditions $x_i(0)$ of all the agents [1], as depicted in **Figure 1b**. In other words, we have

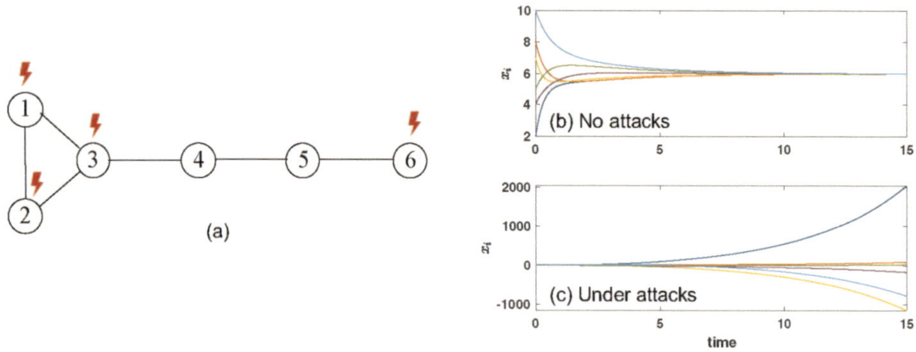

Figure 1.
(a) Network topology of MACS consisting of six agents; (b) under dynamics (8) and in the absence of attacks (d = 0), the states x_i reach average consensus given in (5); (c) The attacker is able to prevent the agents from reaching consensus by launching FDI attacks given in (9) and modifying information sent from agents 1, 2, 3, and 6.

$$x \to x_{ave}\mathbf{1}_n = \underbrace{\left(\frac{1}{n}\sum_{i=1}^{n} x_i(0)\right)}_{x_{ave}}\mathbf{1}_n. \tag{5}$$

Remark 1. The work [16] presents a systematic design for general classes of nonlinear heterogeneous cooperative systems. Specifically, if the individual dynamics can become input passivity-short by a local feedback controller, the dynamic behaviors of the agents at the network level as well as their network control design can then be transformed to (3). Hence, the design presented in the book chapter has broad implications for general cooperative systems.

3.1 System identification toward defending against cyber-attacks on MACS

In practice, the communication network is vulnerable to cyber-attacks. Specifically, this book chapter considers the following attacks: (i) false data injection (FDI) attacks which destroy data integrity of the cooperative system by injecting malicious signals into the communication channels to alter exchanged data; and (ii) denial-of-service (DoS) attacks which block the communication channels causing unavailability of data.

Under DoS attacks, the system matrix $A = [a_{ij}]$ (i.e., Laplacian matrix $-L$) in (4) changes, and the cooperative system may lose its consensus. While agent i can locally estimate the i-th row of the matrix A (which corresponds to the information it receives from its direct neighbors) using the approach presented in Section 6, cooperative stability and performance of MACS are determined by connectivity, collective property of system parameters rather than any specific parameter. There are two ways to identify and thwart DoS attacks. One way is to identify all the system parameter changes in a_{ij} and then attempt to maintain or recover all the original values of a_{ij}. The challenge lies in developing a distributed algorithm so that the individual agent can estimate all the parameters of the matrix A. To this end, each agent can perform the following consensus algorithm

$$\dot{\hat{A}}_i = \sum_{j\in\mathcal{N}_i}\left(\hat{A}_j - \hat{A}_i\right). \tag{6}$$

Here, matrix $\hat{A}_i$ denotes the agent i's a local estimate of the matrix A with

$$\hat{A}_i(0) = n\begin{bmatrix} \mathbf{0}_n^T \\ \vdots \\ \mathbf{0}_n^T \\ [a_{i*}^T] \\ \mathbf{0}_n^T \\ \vdots \end{bmatrix} \rightarrow i\text{-th row}$$

where $[a_{i*}] \in \mathbb{R}^n$ denotes the i-th row of the matrix A. The agents local estimate will then converge to the matrix A given that the communication network's topology is connected. However, this approach has several limitations. First, it is expensive as each agent has to n^2 exchange number of information with its neighbors. Furthermore, the network's connectivity may have already been destroyed before each agent can estimate the overall system parameters of A. To overcome the limitation of the approach described previously and to quantify the connectivity property of a network, let us define the path p_{ij} between node i and node j as

$$p_{ij} = \prod_{k_1, k_2, \cdots, k_l} a_{ik_1} a_{k_1 k_2} \cdots a_{k_l j}, \tag{7}$$

where the sequence $k_1, k_2, \cdots, k_l$ of represents the set of nodes corresponding to the sequence of edges $(i, k_1), (k_1, k_2), \cdots, (k_l, j)$ which connect the two nodes. Since the original graph is connected, the node i knows that $p_{ij} = 1$ for any choice of j. Hence, the second way of distributively detecting DoS attacks and making the system resilient to such attacks is to distributively monitoring/estimating p_{ij} in real-time and ensuring that, at any time, there are at least two distinct paths (or existence of a loop/cycle) between nodes i and j (such that any DoS attack will not destroy the connectivity). Accordingly, one focus of this book chapter is to develop such a distributed algorithm for node i to identify/monitor p_{ij} and maintain the existence of two distinct $p_{ij} = 1$ for any of the node i's neighboring node $j \in \mathcal{N}_i$. As will be discussed in more detail later, this approach only requires each agent to exchange n number of information with its neighbors.

In the following subsections, the vulnerability of the MACS under the consensus algorithm (4) against both attack vectors will be analyzed in more detail.

3.2 Vulnerability of MACS under FDI attacks

The consensus dynamics (4) under unknown FDI attacks can be written as

$$\dot{x} = -L(x - d), \tag{8}$$

where $d(t) \in \mathbb{R}^n$ is an unknown injection launched by the attacker. Specifically, $d_i(t) \neq 0$ means that the attacker modifies the information sent by the agent to its neighbors by injecting a malicious signal $d_i(t)$ and also corrupts the local state feedback of the agent i. It is assumed that the attacker has knowledge of the network topology of the MACS, that is the Laplacian matrix L, and also have access to the physical state x which can used to launch the exogenous injections d. In addition, it is also assumed that the unknown injection d is bounded. This assumption is a reasonable precaution for intelligent attackers to avoid detection. Furthermore, it also aligns with practical considerations, as physical signals, such as those in power systems, typically have a known range. Note that unbounded injections can then be readily rejected using a threshold check [17].

There are many possible choices of d which can prevent the MACS from reaching an average consensus. For example, the attacker can generate the injection d according to the following dynamics, unknown to the defender:

$$\dot{d} = F_a d + B_a x, \tag{9}$$

where matrix F_a is Hurwitz, and together with matrix B_a are designed by the attacker to ensure that one of the eigenvalues of the matrix $\begin{bmatrix} -L & L \\ B_a & F_a \end{bmatrix}$ has a positive real part as illustrated in **Figure 1c**.

3.3 Vulnerability of MACS under DoS attacks

Unlike FDI attacks that compromise data integrity, DoS attacks disrupt the availability of communication channels, effectively isolating agents and impeding their ability to reach consensus. During a DoS attack, malicious actors flood the network's communication channels with overwhelming traffic, rendering them inoperable. This disruption in communication can be modeled by intermittently adjusting the Laplacian matrix $L(t)$ of the consensus algorithm to reflect the compromised links. Consequently, the consensus dynamics under DoS attacks can be expressed as:

$$\dot{x} = A(t) = -L(t)x, \tag{10}$$

where $L(t)$ dynamically changes based on the network's connectivity at the time t, showcasing the impact of DoS attacks.

The resilience of MACS against DoS attacks is significantly influenced by the network's structural properties. The initial impact of a DoS attack on a well-connected MACS might not immediately lead to a disconnection of the network. However, persistent or intense DoS attacks can gradually affect the network's connectivity, leading to the progressive disconnection of $\mathcal{G}$. This disconnection causes the standard consensus algorithm (4) to fail in achieving a unified state among agents.

Critical edges within the communication graph $\mathcal{G}$ play a pivotal role in this context. These edges, if severed, can dramatically impact the network's functionality by partitioning the system into disjoint segments. A critical edge, or bridge, in an undirected network, is defined as an edge whose removal leads to the disconnection of the network, thereby identifying it as a singular point of vulnerability within the network's topology [18]. Borrowing the notation in (7), an edge (i,j) is critical means that there is only one distinct path with $p_{ij} = 1$.

Edge connectivity, on the other hand, is a fundamental metric of network robustness, quantifying the minimum number of edges that must be removed to render the network disconnected. The presence of critical edges directly impacts a network's edge connectivity; specifically, a network with even a single critical edge has an edge connectivity of 1. A network with high-edge connectivity is more robust and capable of withstanding multiple edge failures without losing overall connectivity. In contrast, networks with critical edges are vulnerable, as the failure of just one such edge can compromise network integrity [19].

For example, in **Figure 1a**, the edges between agent pairs (3,4), (4,5), and (5,6) are critical, and it causes the edge connectivity of the network connecting MACS to be one. It follows that identifying and eliminating the criticality of these edges enhances the robustness of MACS and ensures its resilience to disruptions such as failures or targeted DoS attacks.

4. Resilient multi-agent cooperative system against DoS attacks

In what follows, two methods are introduced to robustify MACS against DoS attacks. The first is to make MACS robust by applying the algorithm in Ref. [20] of identifying critical edges within the network's communication graph $\mathcal{G}$ as a result of DoS attacks so that a network reconfiguration could be done to ensure reliable consensus among agents. The second is to introduce a digital twin of the MACS to achieve robustness.

4.1 Distributed algorithms for network reconfiguration

The reconfiguration process is aided by two distributed algorithms, each fulfilling a specific role. The first algorithm is used by the individual agent i to determine its neighbors, and the second is to determine if any of the edges $(i,j), j \in \mathcal{N}_i$, is critical. Upon identifying the critical edge(s), additional edge(s) could be added to ensure resilience against DoS attacks.

Consider two key state vectors for each agent $i \in \mathcal{N}$: $\xi_i(k) = \left[\xi_{i,1}(k) \ \cdots \ \xi_{i,n}(k)\right]^T \in \mathbb{R}^n$ denoting the agent i's estimate of its neighbors, and $\omega_i(k) = \left[\omega_{i,1}(k) \ \cdots \ \omega_{i,n}(k)\right]^T \in \mathbb{R}^n$ denoting agent i's estimate of its neighbor structure. First, each agent i executes the following distributed maximum and minimum protocols for n consecutive iterations, that is $k = 0, \cdots, (n-1)$:

$$\xi_{i,j}(k+1) = \max_{l \in \mathcal{N}_i \cup i} \xi_{l,j}(k), \quad j \in \mathcal{N}, \tag{11}$$

$$\omega_{i,j}(k+1) = \begin{cases} \omega_{i,j}(k) & \text{if } \ \xi_{i,j}(k+1) = \xi_{i,j}(k), \\ \min\limits_{l \in \mathcal{N}_i} \left(\omega_{l,j}(k) + 1\right) & \text{if } \ \xi_{i,j}(k+1) > \xi_{i,j}(k) \end{cases} \tag{12}$$

where the initial values are set to

$$\xi_{i,j}(0) = \begin{cases} 1, & \text{if } j = i \\ 0, & \text{otherwise} \end{cases}; \qquad \omega_{i,j}(0) = \begin{cases} 0, & \text{if } j = i \\ \infty, & \text{otherwise} \end{cases} \tag{13}$$

That is, at the initialization step $k = 0$, agent i is aware only of itself, setting $\xi_{i,i}(0) = 1$ and all other $\xi_{i,j}(0) = 0$, and similarly, $\omega_{i,i}(0) = 0$ with all other $\omega_{i,j}(0)$ set to infinity. In subsequent iterations, the agent i updates its state based on the reachability of other agents. The state $\xi_{i,j}(k+1)$ turns to 1 when agent j becomes reachable within $k+1$ steps from agent i through any neighbor l. Concurrently, the algorithm computes the shortest path $\omega_{i,j}(k+1)$, that is the minimum of $\omega_{l,j}(k) + 1$

across all neighbors $l \in \mathcal{N}_i$ that can reach j whenever a new path is found. These values are non-decreasing and only change when new paths are discovered.

Example 1. *Consider the MACS in* **Figure 1a**, *specifically agent 4. The neighbor set of agent 4 is given by $\mathcal{N}_4 = \{3, 5\}$. Initially at $k = 0$,*

$$\xi_4(0) = [0,0,0,1,0,0]^T; \qquad \omega_4(0) = [\infty, \infty, \infty, 0, \infty, \infty]^T.$$

Then, at $k = 1$

$$\xi_4(1) = [0,0,1,1,1,0]^T; \qquad \omega_4(1) = [\infty, \infty, 1,0,1, \infty]^T.$$

Again at $k = 2$,

$$\xi_4(2) = [1,1,1,1,1,1]^T; \qquad \omega_4(2) = [2,2,1,0,1,2]^T.$$

For the next 4 steps, until $k = 6$, ξ_4 and ω_4 remain the same since there are no new neighbors.

Within the next two steps, that is $k = n \cdots (n+2)$, agent i aims to identify whether any of the edges $(i,j), j \in \mathcal{N}_i$, is critical to estimate if there are at least two distinct paths $p_{ij} = 1$ between nodes i and j. The identification process leverages a specific measure, $\Delta_i^{(il)}$ and j, to discern the necessity of each edge for maintaining connectivity.

Measure Definition: The criticality of an edge (i,l) between agents i and l is assessed through:

$$\Delta_i^{(il)} = \left[\Delta_{i,1}^{(il)} \cdots \Delta_{i,n}^{(il)}\right]^T, \qquad \Delta_{i,j}^{(il)} = \omega_{i,j}(n) - \omega_{l,j}(n), \tag{14}$$

where $\Delta_{i,j}^{(il)}$ captures the shortest path discrepancy from any agent j to i and l, highlighting paths bypassing (i,l).

Specifically, in a connected network, the value of $\Delta_{i,j}^{(il)}$ for each pair of neighboring agents i and l, with respect to any other agent j, can only be -1, 0, or 1. A value of -1 indicates that agent j is relatively closer to the agent i than to agent l, and conversely, a value of 1 suggests j agent is nearer to the agent l. A zero value implies that the agent j is equidistant from both agents i and l.

For every agent $j \in \mathcal{N}$ and for all adjacent agents $i' \in \mathcal{N}_i$ and $l' \in \mathcal{N}_l$, an edge (i,l) is **not** a *critical edge*, i.e., there are at least two distinct paths $p_{il} = 1$ (existence of cycle/loop), under these two scenarios:

- If an agent j is equidistant to agents i and l, it indicates the presence of a cycle that does not rely on (i,l). Such an agent's distance measure, $\Delta_{i,j}^{(il)}$, would be zero, demonstrating that (i,l) is part of a cycle and thus non-critical.

- If, for any agent j, there exist neighboring agents i' of i and l' of l such that agent j is closer to both i' and l' than to i and l, it suggests that j is on a path forming a cycle with (i,l). The distance measures, $\Delta_{i,j}^{(ii')}$ and $\Delta_{l,j}^{(ll')}$, would both be positive, confirming the existence of a cycle.

An edge is determined to be critical when no such equidistant agent or cycle patterns are found, indicating the absence of an alternative path. This means that removing this edge would result in a disconnection of the network.

Example 2. *Consider the MACS in* **Figure 1a**, *specifically the edge between agents* 4 *and* 5. *The neighbor sets are* $\mathcal{N}_4 = \{3, 5\}$, $\mathcal{N}_5 = \{4, 6\}$. *Then we have,*

$$\Delta_4^{4,5} = [-1, -1, -1, -1,1,1]^T,$$
$$\Delta_4^{4,3} = [1,1,1, -1, -1, -1]^T,$$
$$\Delta_5^{5,6} = [-1, -1, -1, -1, -1, 1]^T.$$

Since neither of the scenarios for non-critical edges matches, it can be concluded that the edge $(4, 5)$ *is a critical edge and the only path between nodes 4 and 5 is the edge (4,5).*

The next step is executed only by individually each pair of agents associated with a critical edge, say $\{i, l\}$ with $i \in \mathcal{N}_l^c$ and $l \in \mathcal{N}_i^c$. The goal of this step is to identify the most remote pairs of agents adjacent to each critical edge, called *augmentation nodes*, $\{i', l'\} \in \mathcal{N}_i^r$. These augmentation nodes are strategically chosen as the ones farthest from the critical edge (i, l). If there exist multiple agents at the same distance, the ones with the smallest index are chosen, and it reduces the number of edges added.

$$\{i', l'\} \in \mathcal{N}_i^r \text{ if } (16) \text{ or } (17) \text{ or } (18) \text{ is true,} \tag{15}$$

where

$$\begin{cases} \mu_i = |\mathcal{N}_i| > 1 \\ \mu_l = |\mathcal{N}_l| > 1 \end{cases} \text{and} \begin{cases} i \in \mathcal{N}_l^c, \ l \in \mathcal{N}_i^c, \\ \Delta_{i,i'}^{(il)} = \Delta_{l,l'}^{(li)} = -1 \\ \omega_{i,i'} = \max_k \omega_{i,k}, \\ \omega_{l,l'} = \max_k \omega_{l,k}, \end{cases} \tag{16}$$

$$\begin{cases} \mu_i = |\mathcal{N}_i| = 1 \\ \mu_l = |\mathcal{N}_l| > 1 \end{cases} \text{and} \begin{cases} i \in \mathcal{N}_l^c, \ l \in \mathcal{N}_i^c, \\ \Delta_{l,l'}^{(li)} = -1 \\ i' = i, \\ \omega_{l,l'} = \max_k \omega_{l,k}, \end{cases} \tag{17}$$

$$\begin{cases} \mu_i = |\mathcal{N}_i| > 1 \\ \mu_l = |\mathcal{N}_l| = 1 \end{cases} \text{and} \begin{cases} i \in \mathcal{N}_l^c, \ l \in \mathcal{N}_i^c, \\ \Delta_{i,i'}^{(il)} = -1 \\ \omega_{i,i'} = \max_k \omega_{i,k}, \\ l' = l, \end{cases} \tag{18}$$

and $\mathcal{N}^r = \cup_{i \in \mathcal{N}} \mathcal{N}_i^r$ is the *augmentation action set* to be found distributively. Note that $\mu_i = |\mathcal{N}_i| = \mu_l = |\mathcal{N}_l| = 1$ this is the trivial case of a two-agent network and hence is excluded from consideration.

The next $(n - 1)$ steps is to propagate $\mathcal{N}_i^r$ to all the agents so they all have access to $\mathcal{N}^r$ as follows:

$$\mathcal{N}_i^r(k+1) = \cup_{l \in \mathcal{N}_i \cup \{i\}} \mathcal{N}_l^r(k), \tag{19}$$

where $k = (2n+2), \cdots, (3n+1)$, $\cup$, is the union operation of sets containing non-ordering pairs (that is, if $\{i,j\} \subset \mathcal{N}_l^r(k)$, then $\{j,i\} \subset \mathcal{N}_l^r(k)$), and $\mathcal{N}_i^r(3n+2) = \mathcal{N}_i^r$ is given by (15).

And, as the final step, edge addition is accomplished by the pairs of agents identified in $\mathcal{N}^r$ to complete their connection, thereby eliminating critical edges by ensuring there are at least two distinct paths $p_{ij} = 1$ for any pair of nodes $(i,j) \in \mathcal{E}$. As a result, the network's edge connectivity is increased to 2 which enhances the MACS's resilience against DoS attacks. This approach connects distant edges within acyclic parts of the network, specifically, those separated by critical edges, to both preserve network connectivity under attack and minimize the additional connections required. Further details can be found in Ref. [20].

4.2 Resilient multi-agent cooperative system against DoS attacks: a digital twin approach

An alternative approach to removing the critical edges is achieved by interconnecting the MACS with its digital twin Σ_{dt} as illustrated in **Figure 2**. Specifically, the digital twin consists of virtual agents with a similar number to the MACS and each virtual agent maintains a state z_i. The dynamics of the MACS interconnected with its digital twin are designed as

$$\begin{aligned} \dot{x} &= -L(t)x + Lz, \\ \dot{z} &= -Lz - Lx, \end{aligned} \tag{20}$$

where $z = [z_1, \cdots z_n]^T$. In contrast to the physical state x_i, the state of the virtual agent z_i does not have any physical meaning, hence it is also called a virtual state, which makes it less interesting to the attacker. Furthermore, its initial condition $z_i(0)$ can be set to any value. The digital twin can be implemented using cloud computing in combination with software-defined networking [21] to direct traffic in the network. The network of the digital twin, denoted by $\mathcal{G}_v = (\mathcal{V}, \mathcal{E}_v)$, has the same structure as the MACS and we also call it a virtual network. In addition to the physical state x_j, agent j also sends the virtual state z_j and the masked physical state $x_j + z_j$ to its neighbors via the virtual network.

The interconnection of the physical network and its digital twin is a competitive interaction from a game-theoretical point of view. More on this can be found in the

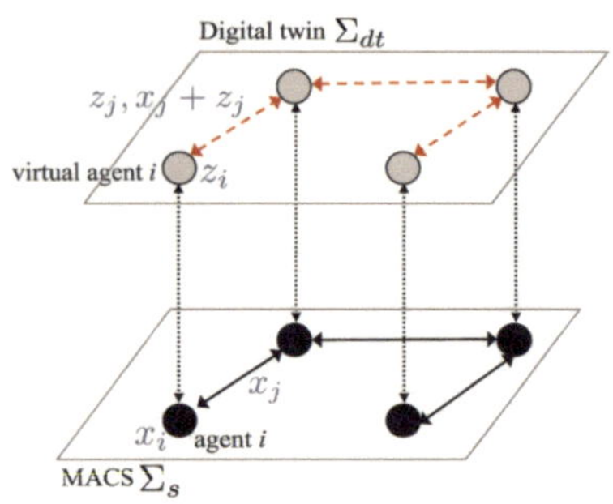

Figure 2.
Interconnection of MACS Σ_s with its digital twin Σ_{dt}. The digital twin consists of the same number of (virtual) agents and network topology as the MACS. Furthermore, each virtual agent maintains a virtual state z_i.

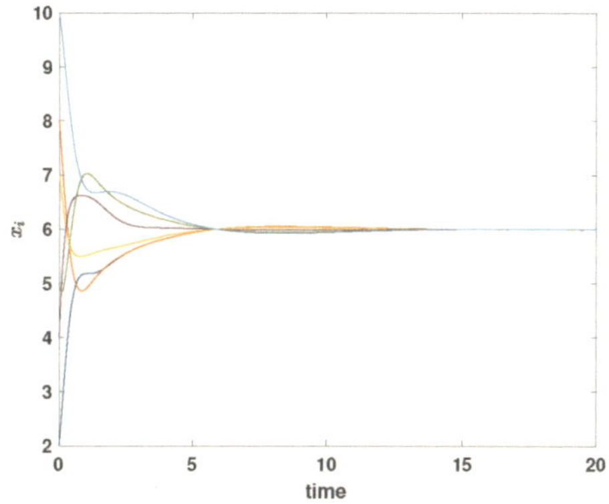

Figure 3.
*Physical states of all the agents converge to the average consensus under interconnected dynamics (20) and with network topology given in **Figure 1a**.*

next section. It is worth noting that the interconnection with the digital twin does not interfere with the operation of the MACS. In other words, under this interconnection, the physical states of the MACS can still be ensured to reach average consensus in (5) as shown in **Figure 3**, refer [22].

If a connected digital twin is introduced, the critical edges are automatically eliminated since the existence of a connected digital twin effectively introduces redundancy in the network's connectivity. This redundancy ensures that even if critical edges in the physical network are severed due to attacks, the interconnected system can maintain operational integrity through the virtual network. This setup leverages the parallelism between the physical and virtual states, allowing the system to sidestep vulnerabilities exposed by the critical edges in the physical network.

Example 3. *Consider a connected MACS with a digital twin where the adversary launches DoS attacks on the network Σ_s. Consequently, we denote the Laplacian matrix of the physical network post-attacks as $\mathcal{L}^a$. Then, for following Lyapunov functions $V(x,z)$,*

$$V = x^T x + z^T z, \tag{21}$$

it follows that under DoS attacks,

$$\begin{aligned} \dot{V} &\leq -x^T L^a x + x^T L_{dt} z - z^T L_{dt} z - z^T L_{dt} x \\ &= -x^T L^a x - z^T L_{dt} z, \end{aligned} \tag{22}$$

where $\mathrm{L}_{\mathrm{dt}} = \mathrm{L}$. *Furthermore, by analyzing the invariant set of $\dot{\mathrm{V}}$ and since the graph $\mathcal{G}_v$ is connected, it can be concluded that the average consensus (5) is ensured under any number of DoS attacks on the physical network Σ_s, as illustrated in* **Figure 4**.

5. Resilient multi-agent cooperative system against FDI attacks: a competitive interaction design via digital twin

The interconnection of the MACS with its digital twin can also be extended to ensure the resilient operation of MACS under FDI attacks. To this end, we introduce a

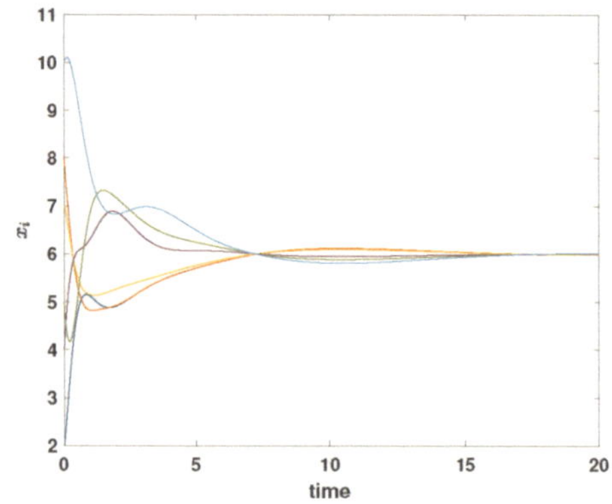

Figure 4.
Physical states of all the agents correspond to the interconnected dynamics (20) and converge to the average consensus under DoS attacks on the physical network Σ_s. Specifically, the adversary disrupts the links (3,4) and (5,6) in ***Figure 1a****, resulting in a disconnected network $\mathcal{G}$.*

scalar gain $\beta > 0$ which represents the strength of the interconnection, resulting in a competitive interaction between the MACS and its digital twin. Hence, the dynamics (20) with the gain β and under unknown FDI attacks can then be written as:

$$\begin{aligned} \dot{x} &= -Lx + \beta Lz + \mathrm{Ld}, \\ \dot{z} &= -Lz - \beta Lx + \mathrm{Ld}', \end{aligned} \tag{23}$$

where it is assumed that the attacker can also modify the information being exchanged via the digital twin's network by inserting a bounded injection d'. Here, agent j sends the information of βz_j and the masked physical state $\beta x_j + z_j$ via the virtual network.

The following results, illustrated in **Figure 5**, can then be obtained [23]:

1. In the absence of FDI attacks, i.e., $d = d' = 0$, the physical states of all the agents reach an average consensus (5).

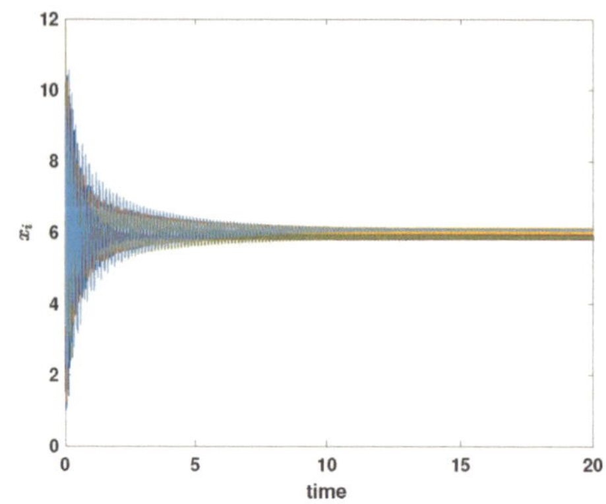

Figure 5.
Physical states of all the agents converge close to the nominal average consensus under interconnected dynamics (23) with $\beta = 150$ and injections d, d' are given in (9). The network topology of MACS is given in ***Figure 1a****.*

2. Under unknown but bounded injections d, d', the physical states of all the agents will converge close to is nominal average consensus value for a sufficiently large value of β, that is

$$\left| \lim_{t \to \infty} x_i(t) - x_{ave} \right| \leq \varepsilon, \quad i = 1, \cdots, n, \tag{24}$$

where ε is a small non-negative constant.

Intuitively, the digital twin acts as an anchor that ensures the resilient operation of the cooperative systems Σ_s for a sufficiently large β. From another perspective, it can be observed that the interactions between the cooperative systems and their digital twin are competitive whose interconnected dynamics (23) correspond to the saddle-point dynamics of a zero-sum game between the two networks [24]. Specifically, the cooperative systems wish to minimize the convex-concave payoff $U = \frac{1}{2}x^T Lx - \frac{1}{2}z^T Lz - \beta x^T Lz$ function by choosing x while its digital twin aims to maximize U by choosing z [24]. Note that the saddle points of U for a sufficiently large β is given by $\mathcal{S} = \text{span}\{(1_n, 0_n), (0_n, 1_n)\}$.

6. Cyber-attack detection and identification

The digital twin not only enhances the resilient of the MACS but also facilitates the detection and identification of cyber-attacks in a distributed manner. Let us denote the possibly corrupted information sent by agent i under dynamics (23) as $\overline{x}_j, \overline{z}_j, \overline{m}_j$ and given by

$$\begin{aligned} \overline{x}_j &= x_j + d_j, \\ \overline{z}_j &= \beta z_j + d_j', \\ \overline{m}_j &= z_j + \beta x_j + d_j'. \end{aligned} \tag{25}$$

where $\overline{x}_j$ is transmitted via the network of the MACS while $\overline{z}_j$ and $\overline{m}_j$ are sent through the virtual network of the digital twin. Using this information, the agent i can detect whether the information that it receives from its neighboring agent is corrupted.

First, let us focus on the detection of DoS attacks. The DoS attack on the link $(i,j) \in \mathcal{E}$ (resp. $(i,j) \in \mathcal{E}_v$) at a certain time results in that agent i does not receive any information from the agent j via the network $\mathcal{G}$ (resp. $\mathcal{G}_v$) and vice versa. Based on this fact, the agent i can detect the DoS attack on a link $(i,j) \in \mathcal{E}$ or $(i,j) \in \mathcal{E}_v$ according to the following simple rules:

1. Link is $(i,j) \in \mathcal{E}$ under DoS attack if and only if $\overline{x}_j = 0$

2. Link $(i,j) \in \mathcal{E}_v$ is under DoS attack if and only if $\overline{z}_j = 0$ and $\overline{m}_j = 0$

Note that the above condition can be distributively checked by the individual agent. Next, we describe how the individual agent can detect the FDI attack on the communication link. The idea is for agent i to estimate the physical state of its neighboring agent x_j using the information that it receives via the virtual network and then to compare the estimated physical state $\hat{x}_j^i$ with the one received via the network of the MACS [25, 26]. To this end, the estimated physical state $\hat{x}_j^i$ by the agent i can be computed from the possibly corrupted information $\overline{z}_j$ and $\overline{m}_j$ in (25) as follow:

$$\hat{x}_j^i = \frac{1}{\beta}\left(\overline{m}_j - \frac{\overline{z}_j}{\beta}\right). \tag{26}$$

Finally, agent i can then detect whether the link $(i,j) \in \mathcal{E}$ or $(i,j) \in \mathcal{E}_v$ is under FDI attacks according to the following rule

1. If the value $|\hat{x}_j^i - \overline{x}_j| > 0$, then agent i can conclude that the communication link $(i,j) \in \mathcal{E}$ and/or $(i,j) \in \mathcal{E}_v$ is under FDI attacks. Note that, however, agent i will not be able to distinguish between the following two cases without any additional information, e.g., if the virtual network is totally secured: the case where one of the links $(i,j) \in \mathcal{E}$ and $(i,j) \in \mathcal{E}_v$ is being attacked and the case where both links in physical and virtual networks are being attacked.

2. If the value $|\hat{x}_j^i - \overline{x}_j| = 0$, then agent i can conclude that there are no FDI attacks on links $(i,j) \in \mathcal{E}$ and/or $(i,j) \in \mathcal{E}_v$ or the attacker successfully launches a stealthy attack. In order to launch stealthy attacks, the attacker must know the structure of virtual states in $\overline{z}_j, \overline{m}_j$ (25) and how they are related to the physical state x_j. However, it is challenging for the attacker to learn these correlations as the gain β is a local information and the virtual state z_i is dynamic (changes over time) and its initial value can be arbitrarily set by the agents.

Figure 6 illustrates how agent 4 detects the FDI attacks on the link (4,3) for the network and attack shown in **Figure 1**. Once the agents identify the compromised communication links, they can then cooperatively remove those links [27] and create new links to the network [28] to ensure its connectivity so that the MACS can recover its nominal average consensus value.

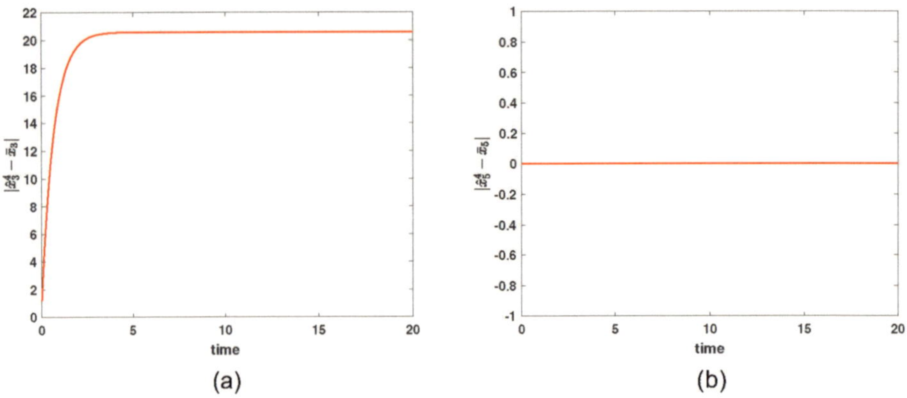

Figure 6.
Attack identification by agent 4 for FDI attacks is shown in ***Figure 1****. By computing the value of* $|\hat{x}_j^i - \overline{x}_j|$*, agent 4 can detect that the communication link (4, 3) is being compromised while the link (4, 5) is not compromised.*

7. Conclusions

The book chapter presents a distributed strategy to enhance the resilience of multi-agent cooperative systems against FDI and DoS attacks. First, a distributed algorithm is proposed to identify critical edges, followed by introducing an adaptive network reconfiguration algorithm to remove those critical edges, thus ensuring resilient operation against arbitrary DoS attacks. A digital twin, interconnected with the cooperative systems, is further designed to increase redundancy of the network and protect the cooperative system from FDI attacks. While this book chapter considers attacks on data integrity and availability, the proposed digital twin is also promising to encounter attacks on data confidentiality [29]. Future work includes the extension of the proposed defense framework to address simultaneous FDI and DoS attacks on both the cooperative systems and its digital twin.

Acknowledgements

The work of Azwirman Gusrialdi was supported by the Research Council of Finland under Academy Project 330073. The work of Zhihua Qu was supported in part by the U.S. Department of Energy's under Award DE-EE0007998, Award DE-EE0009028, Award DE-EE0009152, Award DE-EE0009339, and Award DE-AC05-76RL01830.

Author details

Azwirman Gusrialdi[1*†], Deepalakshmi Babu Venkateswaran[2†] and Zhihua Qu[2]

1 Tampere University, Tampere, Finland

2 University of Central Florida, Orlando, FL, USA

*Address all correspondence to: azwirman.gusrialdi@tuni.fi

† These authors contributed equally.

References

[1] Qu Z. Cooperative Control of Dynamical Systems. London: Springer Verlag; 2009

[2] Wang Y, Yan X, Tang Y, Liao K, Syed MH, Guillo-Sansano E, et al. Aggregated energy storage for power system frequency control: A finite-time consensus approach. IEEE Transactions on Smart Grid. 2018;**10**(4):675-3686

[3] Santini S, Salvi A, Valente AS, Pescapè A, Segata M, Cigno RL. Platooning maneuvers in vehicular networks: A distributed and consensus-based approach. IEEE Transactions on Intelligent Vehicles. 2018;**4**(1):59-72

[4] Aragues R, Cortes J, Sagues C. Distributed consensus on robot networks for dynamically merging feature-based maps. IEEE Transactions on Robotics. 2012;**28**(4):840-854

[5] Savazzi S, Nicoli M, Rampa V. Federated learning with cooperating devices: A consensus approach for massive IoT networks. IEEE Internet of Things Journal. 2020;7(5):4641-4654

[6] Yang T, Yi X, Junfeng W, Yuan Y, Di W, Meng Z, et al. A survey of distributed optimization. Annual Reviews in Control. 2019;**47**:278-305

[7] Gusrialdi A, Zhihua Q. Resilient distributed optimization against cyber-attacks. IEEE Control Systems Letters. 2023;7:3956-3961

[8] Ishii H, Wang Y, Feng S. An overview on multi-agent consensus under adversarial attacks. Annual Reviews in Control. 2022;**53**:252-272

[9] Gusrialdi A, Zhihua Q. Smart grid security: Attacks and defenses. In: Stoustrup J, Annaswamy A, Chakrabortty A, Qu Z, editors. Smart Grid Control: An Overview and Research Opportunities. Switzerland: Springer Verlag; 2018. pp. 199-223

[10] E-ISAC. Defense Use Case. Analysis of the Cyber Attack on the Ukrainian Power Grid. Vol. 388(1-29). Washington, DC: Electricity Information Sharing and Analysis Center (E-ISAC); 2016. p. 3

[11] Dibaji SM, Hussain A, Ishii H. A Tutorial on Security and Privacy Challenges in CPS. Cham: Springer International Publishing; 2022. pp. 121-146

[12] Feng Z, Guoqiang H. Secure cooperative event-triggered control of linear multiagent systems under dos attacks. IEEE Transactions on Control Systems Technology. 2020;**28**(3): 741-752. DOI: 10.1109/TCST.2019.2892032

[13] Zuo Z, Cao X, Wang Y, Zhang W. Resilient consensus of multiagent systems against denial-of-service attacks. IEEE Transactions on Systems, Man, and Cybernetics: Systems. 2022; **52**(4):2664-2675

[14] Usevitch J, Panagou D. Resilient leader-follower consensus to arbitrary reference values in time-varying graphs. IEEE Transactions on Automatic Control. 2020;**65**(4):1755-1762. DOI: 10.1109/TAC.2019.2934954

[15] Eslami A, Abdollahi F, Khorasani K. Stochastic fault and cyber-attack detection and consensus control in multi-agent systems. International Journal of Control. 2022;**95**(9):2379-2397

[16] Zhihua Q, Simaan MA. Modularized design for cooperative control and plug-

and-play operation of networked heterogeneous systems. Automatica. 2014;**50**(9):2405-2414

[17] Gusrialdi A, Zhihua Q, Simaan MA. Competitive interaction design of cooperative systems against attacks. IEEE Transactions on Automatic Control. 2018;**63**(9):3159-3166

[18] Ang-Kun W, Tian L, Liu Y-Y. Bridges in complex networks. Physical Review E. 2018;**97**:012307

[19] Naor D, Gusfield D, Martel C. A fast algorithm for optimally increasing the edge-connectivity. In: Proceedings [1990] 31st Annual Symposium on Foundations of Computer Science. St. Louis, MO: IEEE; Vol. 2. 1990. pp. 698-707. DOI: 10.1109/FSCS.1990.89592

[20] Venkateswaran D, B, Qu Z, Gusrialdi A. A distributed method for detecting critical edges and increasing edge connectivity in undirected networks. In: IEEE 2024 Conference on Decision and Control, Milan, Italy. IEEE; 2024. Submitted

[21] Ala'Darabseh, Freris NM. A software-defined architecture for control of IoT cyberphysical systems. Cluster Computing. 2019;**22**(4):1107-1122

[22] Gusrialdi A, Qu Z, Simaan M. Robust design of cooperative systems against attacks. In: Proceedings of American Control Conference. Portland, OR, USA: IEEE; 2014. pp. 1456-1462

[23] Iqbal M, Zhihua Q, Gusrialdi A. Distributed resilient consensus on general digraphs under cyber-attacks. European Journal of Control. 2022;**68**:100681

[24] Gharesifard B, Başar T. Resilience in consensus dynamics via competitive interconnections. IFAC Proceedings Volumes. 2012;**45**(26):234-239

[25] Gusrialdi A, Iqbal M, Zhihua Q. Towards resilient design of leader-following consensus with attack identification and privacy preservation capabilities. In: European Control Conference. Bucharest, Romania: IEEE; 2023. pp. 1-6

[26] Gusrialdi A, Qu Z. Cooperative systems in presence of cyber attacks: A unified framework for resilient control and attack identification. In: Proceedings of American Control Conference. Atlanta, GA: IEEE; 2022. pp. 330-335

[27] Gusrialdi A. Connectivity-preserving distributed algorithms for removing links in directed networks. Network Science. 2022;**10**(3):215-233

[28] Atman MWS, Gusrialdi A. Finite-time distributed algorithms for verifying and ensuring strong connectivity of directed networks. IEEE Transactions on Network Science and Engineering. 2022;**9**(6):4379-4392

[29] Gusrialdi A. Resilient and privacy-preserving leader-follower consensus in presence of cyber-attacks. IEEE Control Systems Letters. 2023;7:3211-3216

Chapter 9

Innovative Adaptive Imaged Based Visual Servoing Control of 6 DoFs Industrial Robot Manipulators

Rongfei Li and Francis F. Assadian

Abstract

Image-based visual servoing (IBVS) methods have been well developed and used in many applications, especially in pose (position and orientation) alignment. However, most research papers focused on developing control solutions when 3D point features can be detected inside the field of view. This work proposes an innovative feedforward-feedback adaptive control algorithm structure with the Youla parameterization method. A designed feature estimation loop ensures stable and fast motion control when point features are outside the field of view. As 3D point features move inside the field of view, the IBVS feedback loop preserves the precision of the pose at the end of the control period. Also, an adaptive controller is developed in the feedback loop to stabilize the system in the entire range of operations. The nonlinear camera and robot manipulator model is linearized and decoupled online by an adaptive algorithm. The adaptive controller is then computed based on the linearized model evaluated at current linearized point. The proposed solution is robust and easy to implement in different industrial robotic systems. Various scenarios are used in simulations to validate the effectiveness and robust performance of the proposed controller.

Keywords: image-based visual servoing, adaptive control, robust control, feature estimation, feedforward control

1. Introduction

Automatic alignment plays a crucial role in industrial assignments, such as micro-manipulation, autonomous welding, and industrial assembly [1–3]. Visual servoing is a powerful tool that is commonly used in this application and guides the robotic systems to their desired poses [4–6]. The task to be solved by a visual servoing system is to provide velocities of the end-effector that stabilizes and minimizes the difference between image features extracted by a vision device and the desired configurations [7].

The most known visual servoing techniques can be divided into two main categories: Image-based visual servoing (IBVS) and Position-based visual servoing (PBVS) [7]. IBVS is designed to drive the robot manipulators *via* feedback loops that matches

2D image features, such as points [7], lines [8], and entire images [9] while PBVS architecture extracts features in an image and estimates the pose, (including position and orientation) with respect to a 3D coordinate of frame in workspace and the difference between current pose and desired pose is defined as the control error [10]. In either case, an interaction matrix that relates the derivative of an image or pose features to the spatial velocities of the end-effector must be computed [7, 11].

In this research paper, the IBVS structure is explored and applied to a fastening and unfastening manufacturing scenario. Compared to PBVS, IBVS has the advantage of being more robust against camera parametric variations and noise but is more vulnerable to local minima and image singularities [7]. In general, the interaction matrix can be computed by using direct depth information [12, 13] or by approximation of the depth [14, 15] or by depth-independent interaction matrix [16]. However, these methods use redundant features to invert a non-square interaction matrix, as required in controller design, this in turn causes problems of local minima and image singularities. To address these issues, [17] uses the stereo visual system and provides a square interaction matrix. We have seen more trends in using stereo visual configuration in recent visual servoing works [18, 19].

In most related works, a simple kinematic model of the camera is solely used in generating an interaction matrix [7, 20, 21]. Dynamic visual servoing studied by [17, 22, 23] includes the dynamic of robot arms in their models and they argue to have higher performance and enhanced stability in their response. However, controllers built in these papers are usually achieved with the PID control technique or its simplified variations. One improvement in this work is to use Youla robust control design technique [24] that includes both kinematics and dynamics in the model development stage. This advanced control technique can increase the robustness and stability of the system for high-speed tasks.

The eye-in-hand (EIH) vision system and the eye-to-hand (ETH) vision system are two kinds of camera configurations that have been widely used in visual servoing [18, 19]. EIH has the freedom to obtain adequate environmental information, but the camera attached to the robot arms occupies more space and reduces flexibility of robot movement. On the other hand, in ETH systems, robot movements are not affected by the image extraction process. However, it usually suffers information loss and control loss of the robot when the point features are outside the view of the camera. One possible way to tackle this problem is to install several cameras to cover the whole workspace [19, 25]. In this paper, we address this problem by introducing an innovative method of designing a feedforward-feedback control architecture. A designed feature estimation loop ensures stable and fast motion control when point features are outside the field of view. As 3D point features move inside the field of view, IBVS feedback loop preserves the precision of the pose at the end of the control period.

Adaptive control methods have been widely used in recent research on visual servoing fields. Liu and Wang [26] presented a new simplified adaptive controller for visual servoing of robot manipulators, which is based on the Slotine-Li algorithm [27]. They developed an adaptive algorithm to estimate unknown geometric parameters, such as the depths of the image features in the interaction matrix. In addition, in another paper [28], researchers explored a new resilient adaptive dynamic tracking control scheme for a fully uncalibrated IBVS system with unknown actuator faults. An effective adaptive algorithm was developed to estimate uncalibrated parameters in the camera, robot, and end-effector, which appear with a highly coupled and nonlinear form in the composite Jacobian matrix. Moreover, we have seen applications of adaptive controls with neural networks in visual servoings. For example, Qiu and Wu [29] have developed an adaptive neural network-based IBVS dynamic method for both eye-in-hand and eye-to-hand camera configurations with unknown dynamics and external disturbances. In most applications, adaptive algorithms are used to estimate unknown parameters in models. However, adaptive controller can also be used in dealing with coupled and nonlinear dynamic models in the IBVS structures, and this application is less addressed in literature. In this paper, one innovative contribution is to develop an adaptive feedback loop controller based on linear Youla parametrization to stabilize the nonlinear IBVS system in the whole range of operations. Simulation results for the various scenarios are presented and the robustness to noise and model uncertainties in the manufacturing process of fastening and unfastening are examined.

2. System configuration

2.1 System topology for fastening and unfastening scenarios

In this paper, an automatic alignment system is used to align a screwdriver, which is attached to the gripper of the robot arm and guide it to move above a screw at a prescribed location. The system is composed of a 6-DoF robot manipulator, a stereo camera, and the tools as shown in **Figure 1**. In this Figure, a camera is placed on the front of the workspace. The base frame {O} and the end-effector frame {E} are attached to the robot. A cartesian frame {C} is attached to the optical center of the camera. The body of the screwdriver is modeled as a cylinder and its central axis is approximately parallel to the Z-axis of the end-effector frame {E}.

2.2 Point feature extraction

To capture the pose (position and orientation) of the screwdriver, two circular fiducial markers are placed on it. The Hough transformation [30] in computer vision

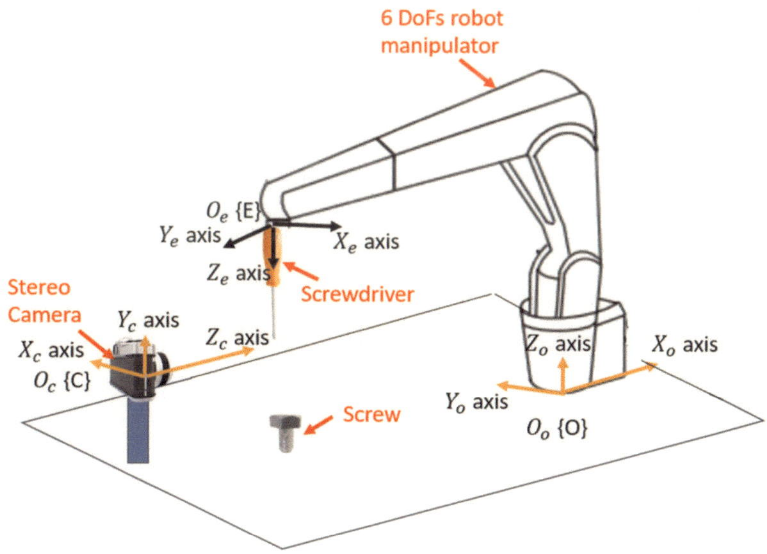

Figure 1.
Configuration of the alignment system.

can isolate circular markers in the image and localize the center of them. A general Hough transformation process is discussed below.

A circle with a radius R and center (a, b) can be defined in the following equation:

$$(x-a)^2 + (y-b)^2 = R^2 \tag{1}$$

Where (x, y) are points on the perimeter of the circle. The three-element tuple (a, b, R) uniquely parameterizes a circle in an image.

To provide an example for this mapping, we only consider one of the circles on our tool and explain the process as follows: the locus of (a, b) points in the parameters space falls on a circle of radius R centered at (x, y), which are mapped to the parameters space and only three points are shown for this example. Each point in a geometric space generates a circle in the parameters space as shown in **Figure 2**. In case the geometric space points belong to the same circle as shown, the center of this circle will be represented by the intersection of circles in the parameters space. During the calculation of the Hough transformation, each image point with the coordinates (x, y) generates tuples (a, b, R) with the unknown R.

By creating different circles in geometric space, we can then generate different tuples. The tuple with the most intersections of the circles, in parameters space, provides the center and the radius of the center in the geometric space.

Then the stereo camera provides the image coordinates of the center of the two fiducial markers as:

$$\widehat{p_{I_1}^T} = \left[\widehat{ul_{I_1}}\ \widehat{ur_{I_1}}\ \widehat{v_{I_1}}\right]^T \tag{2}$$

$$\widehat{p_{I_2}^T} = \left[\widehat{ul_{I_2}}\ \widehat{ur_{I_2}}\ \widehat{v_{I_2}}\right]^T \tag{3}$$

Where $\widehat{p_{I_1}^T}$ and $\widehat{p_{I_2}^T}$ are image coordinates of circular centers of the first and the second fiducial markers. $\widehat{ul}$ and $\widehat{ur}$ are u axis coordinates measured on the left and right lens's plane of the stereo camera, and $\widehat{v}$ is the v axis coordinates.

2.3 Control system block diagram

The feedback-feedforward pose alignment control block diagram with feature estimation feedback is given in **Figure 3**. The point features of fiducial markers extracted from images captured by the stereo camera. The current point features are

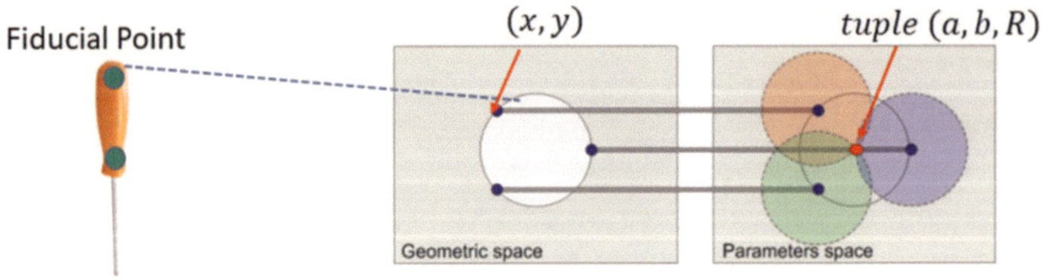

Figure 2.
Feature extraction by Hough transformation. (note: The red point in parameters space represents a possible tuple (a, b, R).

DOI: http://dx.doi.org/10.5772/intechopen.1004857

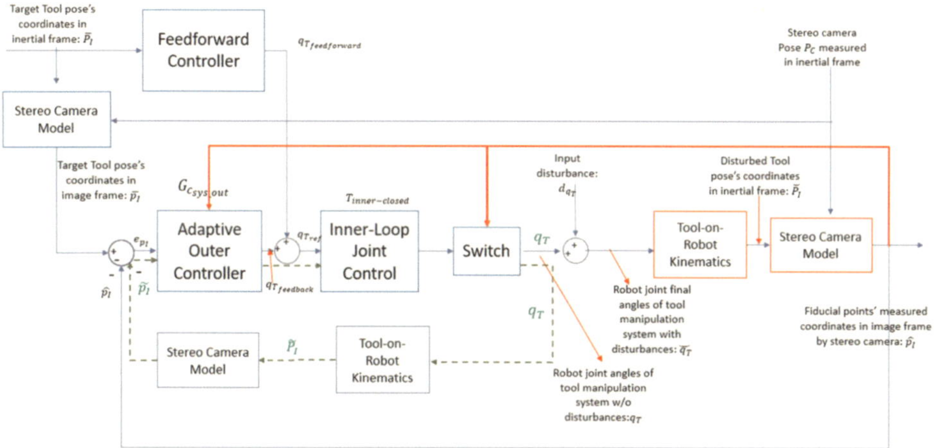

Figure 3.
Block diagram of feedforward-feedback control loop with feature estimation (note: Blocks in red lines are referred to HIL models).

denoted as $\widehat{p_I} = \left[\widehat{p_{I_1}^T}, \widehat{p_{I_2}^T},\right]$, and the target point features are denoted as $\overline{p_I} = \left[\overline{p_{I_1}^T}, \overline{p_{I_2}^T},\right]$. During the alignment, the coordinates of the target point features are fixed in the image frames, while the coordinates of the current point features vary with the movement of the end-effector.

The inner-loop joint control is an inner feedback loop (which is not shown in this Figure, and the detail of it will be discussed in the later section) that moves the robot arms to the target rotational angle q_{Tref}, which is commanded by the outer feedback controller and feedforward controller. Dynamic model of the robot manipulator is included in the inner-loop controlled system. Sensor noise may originate in the inner joint control loop from the low fidelity cheap encoder joint sensors and the dynamic errors from the joint, e.g., compliances. All sources of noise from the joint control loop are combined and modeled as an input disturbance, d_{q_T}, to the outer control loop.

The feedforward control loop is an open loop that brings the tool as close to the target position as possible in the presence of the input disturbance, d_{q_T}. The outputs of the feedforward are the reference joint angles of rotations, $q_{T\_feedforward}$, which are added to the outer feedback controller outputs, $q_{T\_feedback}$, and set as the targets for the joint control inner loop. The function of the Tool-on-Robot kinematics model is to transform a set of current joint angles of the tool manipulator to the current pose of the tool on the end-effector using a kinematic model of the robot arm.

The movement of the tool can be adjusted by the adaptive feedback control loop. The adaptive controller is computed online according to the current image coordinate $\widehat{p_I}$. The feedback control loop rejects the input disturbance, d_{q_T}, by minimizing the error between the target coordinates of two makers, $\overline{p_I}$, and the image estimation, $\widehat{p_I}$ in the image frame.

The feedforward and feedback controllers work simultaneously to move the tool to the target pose location in the tool manipulation system. The combined target $q_{T_{ref}}$ are the inputs to the joint control loop so that both controllers manipulate the tool pose. The benefit of designing both feedback and feedforward controls for the manipulation system is to reduce the time duration. If only feedback control is utilized, the pose estimation generated from the visual system requires image processing and makes the tool movement very slow. We can divide the task of the tool movement control into two stages. In the first stage, under the action of the feedforward control, the tool moves to an approximate location that is close to the desired destination. In the second stage, the feedback controller moves the tool to the precise target location using the tool pose estimation from the camera. In addition, the camera has a range of view and can only detect the tool and measure its 2D feature as $\widehat{p}_I$ when it is not far away from the target. When the tool is moving from a location that is not in the camera range of view, we must estimate the feature as $\tilde{p}_I$ until the tool moves into the range of view.

We developed another feedback loop so that we can estimate the 2D coordinates of the tool from the same mathematical model similar to the Hardware-In-Loop (HIL) models when the real measurements are unavailable. In **Figure 3**, the normal feedback loop (blue lines) is preserved when the tool is inside the camera range of view and hence, the camera can measure the tool's 2D feature $\widehat{p}_I$. However, when the tool is outside the range of view, the 2D feature can only be estimated as $\tilde{p}_I$ (green dashed lines). We can implement a bump-less switch to smoothly switch between these modes of operation. The switching signal is activated when the tool moves in or out of the camera range of view (or when the camera detects fiducial point features).

3. Model development

3.1 Tool-on-robot kinematics

The robotic model we use in this article is a specific manipulator ABB IRB 4600 [31], which is an elbow manipulator with spherical wrist as shown in **Figure 4**. The physical dimension parameters of this robot manipulator $(a_1, L_1, L_2, L_3, L_4, L_t)$ are summarized in the Appendix section. This model of robot has a total of 6 links with three composed of the robot arms and other three composed of wrists. A joint is connected between each of the two adjacent links and there is a total of 5 convolutional joints. In addition to the base of the robot arm, we have attached to each joint a cartesian coordinate, as shown in **Figure 4**. Joint axes Z_0, ... Z_5 are the rotational direction of each joint. Their rotational angles are defined as θ_1, ... θ_6. The benefit of spherical wrist is that the wrist joint center (where Z_3, Z_4 and Z_5 intersect) is kept stationary whichever the wrist orients. Therefore, the task of orientation will not affect the position of the wrist center. Each coordinate attached to the frame is generated based on the procedures that derive the forward kinematics by Denavit-Hartenberg convention (or D-H convention) [32]. In the coordinate frame $O_6X_6Y_6Z_6$ attached to the end-effector, set the unit vector $\widehat{k_6} = \widehat{a}$ along the direction of z_6. Set $\widehat{j_6} = \widehat{s}$ in the direction of gripper closure and set $\widehat{i_6} = \widehat{n}$ as $\widehat{s} \times \widehat{a}$. The values of $\widehat{a}, \widehat{s}$ *and* $\widehat{n}$ in the base frame $O_0X_0Y_0Z_0$ define the orientation of the object (tool) mounted at the end-effector.

In D-H convention, each homogeneous transformation matrix $\boldsymbol{A}_i^{i-1}$ (from frame $\boldsymbol{i}$ to frame $\boldsymbol{i}-\mathbf{1}$) can be represented as a product of four basic transformations:

$$\boldsymbol{A}_i^{i-1} = Rot_{z,\theta_i} Trans_{z,d_i} Trans_{x,a_i} Rot_{x,\alpha_i} \tag{4}$$

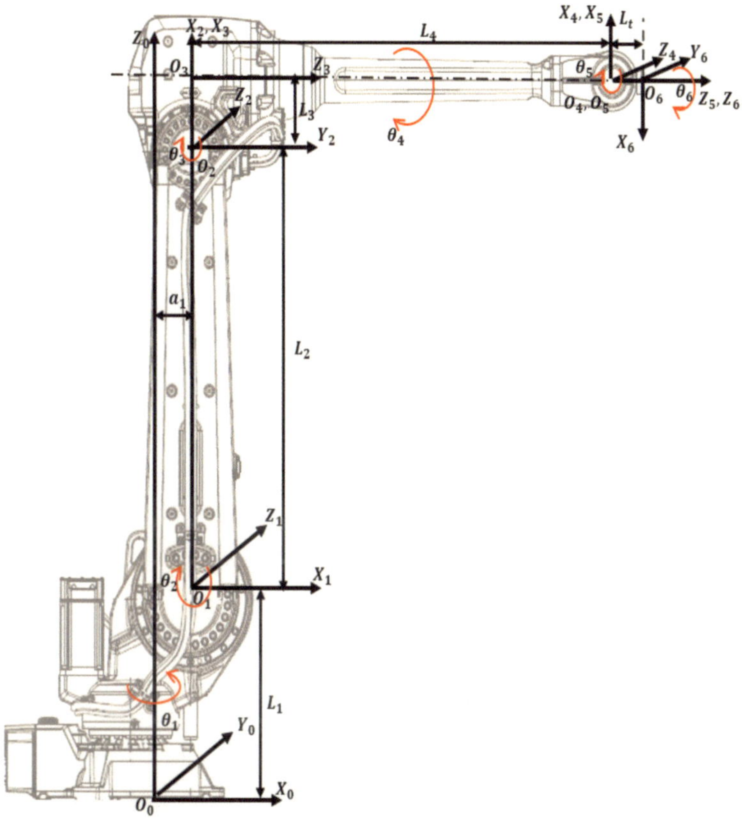

Figure 4.
IRB ABB 4600 model with attached frames.

$$= \begin{bmatrix} c_{\theta_i} & -s_{\theta_i} & 0 & 0 \\ s_{\theta_i} & c_{\theta_i} & 0 & 0 \\ 0 & 0 & 1 & 0 \\ 0 & 0 & 0 & 1 \end{bmatrix} * \begin{bmatrix} 1 & 0 & 0 & 0 \\ 0 & 1 & 0 & 0 \\ 0 & 0 & 1 & d_i \\ 0 & 0 & 0 & 1 \end{bmatrix} * \begin{bmatrix} 1 & 0 & 0 & a_i \\ 0 & 1 & 0 & 0 \\ 0 & 0 & 1 & 0 \\ 0 & 0 & 0 & 1 \end{bmatrix} * \begin{bmatrix} 1 & 0 & 0 & 0 \\ 0 & c_{\alpha_i} & -s_{\alpha_i} & 0 \\ 0 & s_{\alpha_i} & c_{\alpha_i} & 0 \\ 0 & 0 & 0 & 1 \end{bmatrix}$$
$$= \begin{bmatrix} c_{\theta_i} & -s_{\theta_i}c_{\alpha_i} & s_{\theta_i}s_{\alpha_i} & a_i c_{\theta_i} \\ s_{\theta_i} & c_{\theta_i}c_{\alpha_i} & -c_{\theta_i}s_{\alpha_i} & a_i s_{\theta_i} \\ 0 & s_{\alpha_i} & c_{\alpha_i} & d_i \\ 0 & 0 & 0 & 1 \end{bmatrix} \quad (5)$$

Note:

$$c_{\theta_i} \equiv \cos(\theta_i), c_{\alpha_i} \equiv \cos(\alpha_i), s_{\theta_i} \equiv \sin(\theta_i), s_{\alpha_i} \equiv \sin(\alpha_i) \quad (6)$$

Where θ_i, a_i, α_i and d_i are parameters of link i and joint i, a_i is the link length, θ is the rotational angle, α_i is the twist angle and d_i is the offset length between the previous $(i-1)^{th}$ and the current i^{th} robot links. The quantities of each parameter q_i, a_i, α_i and d_i in Eq. (5) are calculated in **Table 1** based on the steps in [32].

Link	a_i	α_i(rad)	d_i	θ_i(rad)
1	a_1	$-\pi/2$	L_1	θ_1^*
2	L_2	0	0	$\theta_2^* - \pi/2$
3	L_3	$-\pi/2$	0	θ_3^*
4	0	$\pi/2$	L_4	θ_4^*
5	0	$-\pi/2$	0	θ_5^*
6	0	0	L_t	$\theta_6^* + \pi$

Only the angle are variables and shown with.*

Table 1.
DH-parameter for elbow manipulator with spherical wrist.

We can generate the transformation matrix $\boldsymbol{T_E^O}$ from the base inertial frame $\boldsymbol{O_0X_0Y_0Z_0}$ $(\boldsymbol{P^o})$ to the end-effector frame $\boldsymbol{O_6X_6Y_6Z_6}$ $(\boldsymbol{P^E})$:

$$T_E^O = A_1^0 A_2^1 A_3^2 A_4^3 A_5^4 A_6^5 \tag{7}$$

Assume a screwdriver with a length of L_{tool} is grasped by the gripper at the end effector, and the body of screwdriver is parallel to the Z_6 axis. Two fiducial feature points attached to it are placed on the screwdriver where their coordinates are measured in the end-effector frame as $\breve{P}_{I_1}\ \{E\} = [0, 0, 0]^T$, and $\breve{P}_{I_2}\ \{E\} = [0, 0, L_{tool}/2]^T$. Then those coordinates of points in the frame $\boldsymbol{P^E}$ can be transformed to the frame $\boldsymbol{P^o}$ (as)

$$\breve{P}_{I_1}\ \{O\} = T_E^O\ \breve{P}_{I_1}\ \{E\} \tag{8}$$

$$\breve{P}_{I_2}\ \{O\} = T_E^O\ \breve{P}_{I_2}\ \{E\} \tag{9}$$

To be consistent with notations in the control block diagram (**Figure 3**), the disturbed angles of rotation in transformation matrix T_E^O should be written as $\breve{q}_T = [\theta_1^*, \theta_2^*, \theta_3^*, \theta_4^*, \theta_5^*, \theta_6^*]^T$.

Take $\breve{P}_I\ \{E\} = \left[\breve{P}_{I_1}\ \{E\}, \breve{P}_{I_2}\ \{E\}\right]$, $\breve{P}_I\ \{O\} = \left[\breve{P}_{I_1}\ \{O\}, \breve{P}_{I_2}\ \{O\}\right]$, the kinematics model can be written as a nonlinear function as follows,

$$\breve{P}_I\ \{O\} = T_E^O\left(\breve{q}_T\right) \cdot \breve{P}_I\ \{E\} \tag{10}$$

3.2 Dynamic models of 6 DoFs robot manipulator

Dynamic models are included in the inner control loop in **Figure 3**. The dynamic model of a serial of 6-link rigid, non-redundant, fully actuated robot manipulator can be written as:

$$(D(q) + J)\ddot{q} + \left(C(q, \dot{q}) + \frac{B}{r}\right)\dot{q} + g(q) = u \tag{11}$$

Where $q \epsilon \mathbb{R}^{6X1}$ is the vector of joint positions, and $u \epsilon \mathbb{R}^{6X1}$ is the vector of electrical power input from DC motors inside joints, $D(q) \epsilon \mathbb{R}^{6X6}$ is the symmetric positive defined matrix, $C(\boldsymbol{q}, \dot{\boldsymbol{q}}) \epsilon \mathbb{R}^{6X6}$ is the vector of centripetal and Coriolis effects, $g(\boldsymbol{q}) \epsilon \mathbb{R}^{6X1}$ is the vector of gravitational torques, $J \epsilon \mathbb{R}^{6X6}$ is a diagonal matrix expressing the sum of actuator and gear inertias, $B \epsilon \mathbb{R}^{6X1}$ is the damping factor, $r \epsilon \mathbb{R}^{6X1}$ is the gear ratio.

3.3 Stereo camera model

Assume the camera's position and orientation is fixed and pre-known in the inertial frame. The transformation matrix T_C^O from the camera attached cartesian frame {C} to the inertial base frame {O} is:

$$T_C^O = \begin{bmatrix} n_x & s_x & a_x & d_x \\ n_y & s_y & a_y & d_y \\ n_z & s_z & a_z & d_z \\ 0 & 0 & 0 & 1 \end{bmatrix} \tag{12}$$

Where $[n_x, n_y, n_z]^T$, $[s_x, s_y, s_z]^T$ and $[a_x, a_y, a_z]^T$ are the camera's directional vector of yaw, pitch, and roll in the base frame $O_0X_0Y_0Z_0$. And $[d_x, d_y, d_z]^T$ are the vector of absolute position of the center of the camera in the base frame $O_0X_0Y_0Z_0$.

The transformation matrix T_O^C from the inertial base frame {O} to the camera frame {C} can be derived as:

$$T_O^C = T_C^{O^{-1}} \tag{13}$$

Then the coordinates of fiducial points in frame $\{O\}$ as expressed in Eqs. (8) and (9) can be transformed to the frame {C} as:

$$\breve{P}_{I_1}\ \{C\} = T_O^C\ \breve{P}_{I_1}\ \{O\} \tag{14}$$

$$\breve{P}_{I_2}\ \{C\} = T_O^C\ \breve{P}_{I_2}\ \{O\} \tag{15}$$

In order to detect depth of an object, a stereo camera is required for implementation. As shown in **Figure 5**, two identical cameras are separated by a baseline distance b. In this paper, the pinhole camera model is used to represent each camera. A 3D point $\breve{P}_I\ \{C\} = [X_I^C, Y_I^C, Z_I^C]^T$, measured in the camera cartesian frame {C}, is projected to two parallel virtual image planes, and each plane is located between each optical center (C_l or C_R) and the object point $\breve{P}_I$. Denote the coordinates of the point projected on the left image plane as $[ul\ v]^T$ and the coordinates of the point projected on the right image plane as $[ur\ v]^T$.

The relationship between the image coordinates $(ul\ v)$, $(ur\ v)$, and the coordinates (X_I^C, Y_I^C, Z_I^C) on the normalized imaging plane is given by:

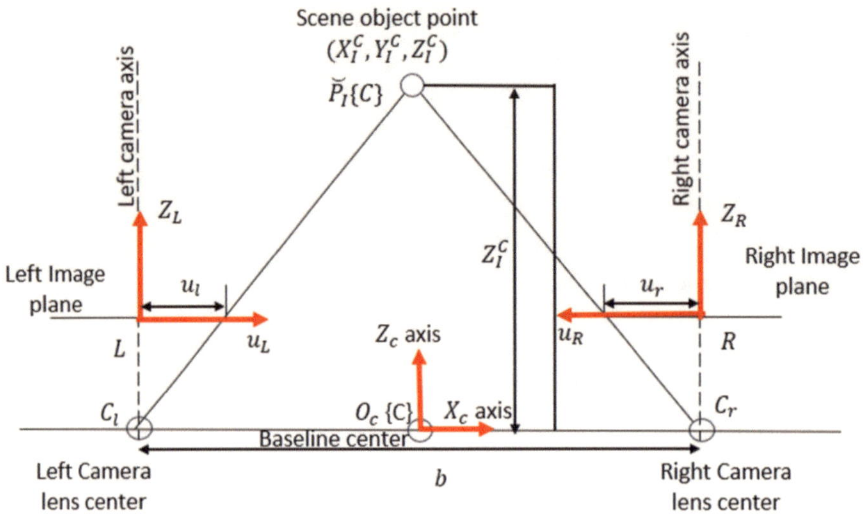

Figure 5.
The projection of a scene object on the stereo camera's image planes. Note: v coordinate on each image plane is not shown in the plot but is measured along the axis that is perpendicular to and point out of the plot.

$$\begin{bmatrix} u_l \\ v_l \\ 1 \end{bmatrix} = \frac{1}{Z^C} \begin{bmatrix} f_u & s_c & u_0 \\ 0 & f_v & v_0 \\ 0 & 0 & 1 \end{bmatrix} \begin{bmatrix} X_I^C \\ Y_I^C \\ Z_I^C \end{bmatrix} - \frac{b}{2Z^C} \begin{bmatrix} f_u \\ 0 \\ 0 \end{bmatrix} \tag{16}$$

$$\begin{bmatrix} u_r \\ v_r \\ 1 \end{bmatrix} = \frac{1}{Z^C} \begin{bmatrix} f_u & s_c & u_0 \\ 0 & f_v & v_0 \\ 0 & 0 & 1 \end{bmatrix} \begin{bmatrix} X_I^C \\ Y_I^C \\ Z_I^C \end{bmatrix} + \frac{b}{2Z^C} \begin{bmatrix} f_u \\ 0 \\ 0 \end{bmatrix} \tag{17}$$

Where $\boldsymbol{f_u}$ and $\boldsymbol{f_v}$ are the horizontal and the vertical focal lengths, and $\boldsymbol{s_c}$ is a skew coefficient. In most cases, $\boldsymbol{f_u}$ and $\boldsymbol{f_v}$ are different if the image horizontal and vertical axes are not perpendicular. In order not to have negative pixel coordinates, the origin of the image plane will be usually chosen at the upper left corner instead of the center. $\boldsymbol{u_0}$ and $\boldsymbol{v_0}$ describe the coordinate offsets.

When there are no coordinate offsets and skews between u-v image plane, that is $s_c = \boldsymbol{u_0} = \boldsymbol{v_0} = \boldsymbol{0}$, the above relationships between 3D domain to 2D image domains of a point $\boldsymbol{P}$ in the workspace can be simplified as:

From 3D to 2D:

$$v = v_l = v_r = \frac{\boldsymbol{Y_I^C}}{Z_I^C} f_v \tag{18}$$

$$u_l = \frac{2\boldsymbol{X_I^C} - b}{2Z_I^C} f_u \tag{19}$$

$$u_r = \frac{2\boldsymbol{X_I^C} + b}{2Z_I^C} f_u \tag{20}$$

Note v_l and v_r have the same value and they are denoted as the same parameter v.

Finally, from stereo camera model equations above, the coordinates of two fiducial points in the image frame can be expressed as:

$$\widehat{p_{I_1}} = \left[\frac{2X_{I_1}^C - b}{2Z_{I_1}^C} f_u, \frac{2X_{I_1}^C + b}{2Z_{I_1}^C} f_u, \frac{Y_{I_1}^C}{Z_{I_1}^C} f_v\right]^T = [ul_{I_1}\ ur_{I_1}\ v_{I_1}]^T \tag{21}$$

$$\widehat{p_{I_2}} = \left[\frac{2X_{I_2}^C - b}{2Z_{I_2}^C} f_u, \frac{2X_{I_2}^C + b}{2Z_{I_2}^C} f_u, \frac{Y_{I_2}^C}{Z_{I_2}^C} f_v\right]^T = [ul_{I_2}\ ur_{I_2}\ v_{I_2}]^T \tag{22}$$

Where $\breve{P}_{I_1}\ \{C\} = \left[X_{I_1}^C, Y_{I_1}^C, Z_{I_1}^C\right]$, and $\breve{P}_{I_2}\ \{C\} = \left[X_{I_2}^C, Y_{I_2}^C, Z_{I_2}^C\right]$ are coordinates of two fiducial points measured in the camera's frame {C}.

A specific stereo camera model: Zed 2 [33] is used in the model development and simulations. The parameters of this type are summarized in the Appendix.

4. Control strategy

4.1 Inner-loop joint control

Our previous paper [34] has provided details of developing the inner-loop joint controller with Youla feedback linearization. Here in this section, we only discuss a summary of the design.

Eq. (11) expresses the dynamic model of a 6 DoFs robot manipulator. We simplify this equation as follows:

$$(q)\ddot{q} + h(q, \dot{q}\ ) = \mathrm{u} \tag{23}$$

$$\text{with } M(q) = D(q) + J \tag{24}$$

$$h(q, \dot{q}\ ) = \left(C(q, \dot{q}) + \frac{B}{r}\right)\dot{q} + g(q) \tag{25}$$

Then, transform the control input as following:

$$\mathrm{u} = M(q)v + h(q, \dot{q}\ ) \tag{26}$$

where v is a virtual input. Then, substitute for u in Eq. (23) using Eq. (26), and since $M(q)$ is invertible, we will have a reduced system equation as follows:

$$\ddot{q} = v \tag{27}$$

By using feedback linearization with Youla parameterization, we can design the controller system as shown in **Figure 6**.

The Joint controller is designed as:

$$G_{C_{sys\_inner}} = \frac{3\tau_{in}{}^2 s + 1}{\tau_{in}{}^3 s + 3\tau_{in}{}^2} \cdot I_{6\times 6} \tag{28}$$

And the closed-loop transfer function can be expressed as:

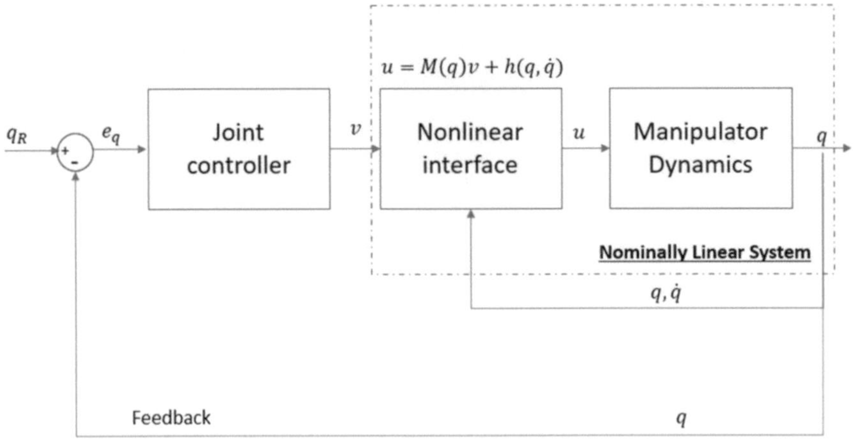

Figure 6.
Block diagram of inner joint control loop with feedback linearization.

$$T_{sys\_inner} = \frac{(3\tau_{in}s + 1)}{(\tau_{in}s + 1)^3} \cdot I_{6\times6} \tag{29}$$

Where $I_{6\times6}$ is a 6x6 identity matrix, and τ_{in} specifies the pole and zero locations and represents the bandwidth of the control system. τ_{in} can be tuned so that the response can be fast with less-overshoot.

4.2 Adaptive controller design

In control system block diagram (**Figure 3**), the linear closed-loop transfer function of inner-loop joint control is derived in Eq. (29). The tool-on-robot kinematics model and stereo camera model, which can be combined as a new model: camera-and-robot model, compose of the nonlinear part of the plant in outer feedback loop. Combining the equations in Section 3, we can derive a nonlinear function denoted as $\mathcal{F}$ that relates the disturbed robot joint angles $\breve{q}_T$ and 2D image coordinates of two fiducial points $\widehat{p}_I$:

$$\widehat{p}_I = \mathcal{F}\left(\breve{q}_T\right) \tag{30}$$

We can use model linearization method by linearizing the nonlinear function in Eq. (30) at different linearized points and design linear controllers utilizing these linearized models. Choosing a linearized point $\breve{q}_T\ 0$, the nonlinear function in Eq. (30) can be linearized using the Jacobian matrix form as:

$$\widehat{p}_I = J\left(\breve{q}_T\ 0\right)\breve{q}_T + \mathcal{F}\left(\breve{q}_T\ 0\right) \tag{31}$$

Where $J\left(\breve{q}_T\ 0\right) \epsilon\, \mathbb{R}^{6X6}$ is the Jacobian matrix of $\mathcal{F}\left(\breve{q}_T\right)$ evaluated as $\breve{q}_T = \breve{q}_T\ 0$.

Assuming $C_1 = J\left(\breve{q}_T\ 0\right)$, $C_2 = \mathcal{F}\left(\breve{q}_T\ 0\right)$, therefore, Eq. (31) can be rewritten as:

$$\widehat{p}_I = C_1 \breve{q}_T + C_2 \tag{32}$$

Let us define $\widehat{p}_I{}' = \widehat{p}_I - C_2$, then, the overall block diagram of the linearized system is shown in **Figure 7**.

The linearized plant transfer function is derived as:

$$G_{p_{sys\_outer}}^{\ linear} = \frac{\widehat{p}_I{}'}{q_{V_{ref_1}}} = C_1 \frac{(3\tau_{in}s + 1)}{(\tau_{in}s + 1)^3} \cdot I_{6\times 6} \tag{33}$$

As C_1 is coupled, the first step to derive a controller for the multivariable system using model linearization is to find the Smith-McMillan form of the plant [35].

To obtain the Smith-McMillan form, we can decompose G_p using singular value decomposition (SVD) as:

$$G_{p_{sys\_outer}}^{\ linear} = U_L M_p U_r \tag{34}$$

where $U_L \epsilon\ \mathbb{R}^{6X6}$ and $U_r\ \epsilon\ \mathbb{R}^{6X6}$are the left and right unimodular matrices, and $M_p\ \epsilon\ \mathbb{R}^{6X6}$ is the Smith-McMillan form of $G_{p_{sys\_outer}}^{\ linear}$. The SVD in this case only works for the replacement of Smith-McMillan form computation because we are dealing with a constant matrix and a dialogized transfer function matrix.

M_p is a diagonalized transfer function matrix with each nonzero entry equal to a gain multiple the transfer function $\frac{(3\tau_{in}s+1)}{(\tau_{in}s+1)^3}$; For the ith row of M_p the entry on the diagonal is:

$$M_p(i,i) = gain(i) \cdot \frac{(3\tau_{in}s + 1)}{(\tau_{in}s + 1)^3}, i\epsilon(1,2,3,4,5,6) \tag{35}$$

Where $gain\ \epsilon\ \mathbb{R}^{6X1}$ is a numerical vector.

The design of a Youla controller for each nonzero entry in M_p is trivial in this case as all poles/zeros of the plant transfer function matrix are in the left half-plane, and therefore, they are stable. In this case, the selected decoupled Youla transfer function matrix, M_Y, can be selected to shape the decoupled closed-loop transfer function matrix, M_T. All poles and zeros in the original plant can be canceled out and new poles and zeros can be added to shape the closed-loop system. Let us select a Youla transfer function matrix so that the decoupled closed-loop SISO system behaves like a second-order Butterworth filter, such that:

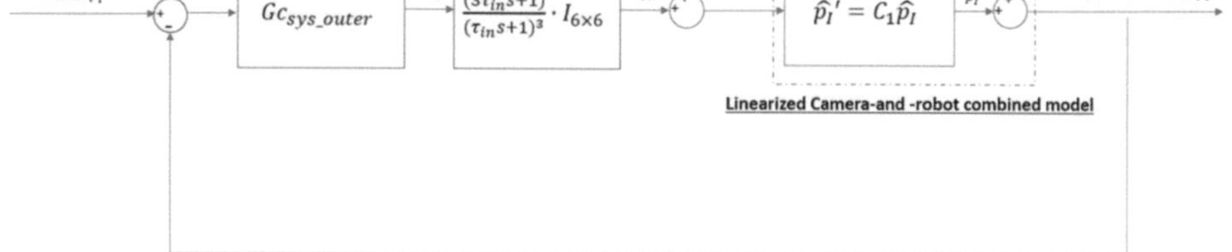

Figure 7.
Block diagram of feedback loop with linearized camera-and-robot combined model.

$$M_T = \frac{{\omega_n}^2}{(s^2 + 2\zeta\omega_n s + {\omega_n}^2)} \cdot I_{6\times 6} \tag{36}$$

where ω_n is called natural frequency and approximately sets the bandwidth of the closed-loop system. It must be ensured that the bandwidth of the outer-loop is smaller than the inner-loop, i.e., $1/\omega_n > \tau_{in}$. Parameter, ζ, is called the damping ratio, which is another tuning parameter.

We can then compute the decoupled diagonalized Youla transfer function matrix M_Y.The diagonal entry of ith row is denoted as $M_Y(i,i)$:

$$M_Y(i,i) = \frac{M_T(i,i)}{M_p(i,i)} = \frac{1}{gain(i)} \frac{{\omega_n}^2}{(s^2 + 2\zeta\omega_n s + {\omega_n}^2)} \frac{(\tau_{in}s+1)^3}{(3\tau_{in}s+1)}, , i\epsilon(1,2,3,4,5,6) \tag{37}$$

The final coupled Youla, closed loop, sensitivity, and controller transfer function matrices are computed as:

$$Y^{linear}_{sys\_outer} = U_R M_Y U_L \tag{38}$$

$$T_{y_{sys\_outer}}^{\ linear} = G_{P_{sys\_outer}}^{\ linear} \cdot Y^{linear}_{sys\_outer} \tag{39}$$

$$S_{y_{sys\_outer}}^{\ linear} = 1 - T_{y_{sys\_outer}}^{\ linear} \tag{40}$$

$$G_{C_{sys\_outer}}^{\ linear} = Y^{linear}_{sys\_outer} \cdot \left(S_{y_{sys\_outer}}^{\ linear}\right)^{-1} \tag{41}$$

The controller developed in the above section is based on the linearization of the combined model at a particular linearized point $\breve{q}_T\ 0$. This controller can only stabilize at a certain range of joint angles around $\breve{q}_T\ 0$. As current joint angles $\breve{q}_T$ deviates from $\breve{q}_T\ 0$, the error between the estimated linearized system in Eq. (33) and the true nonlinear system in Eq. (30) increases.

To tackle this problem, we develop an adaptive controller that is computed online based on linearization of the model at current joint angles. This control process is depicted in **Figure 8**.

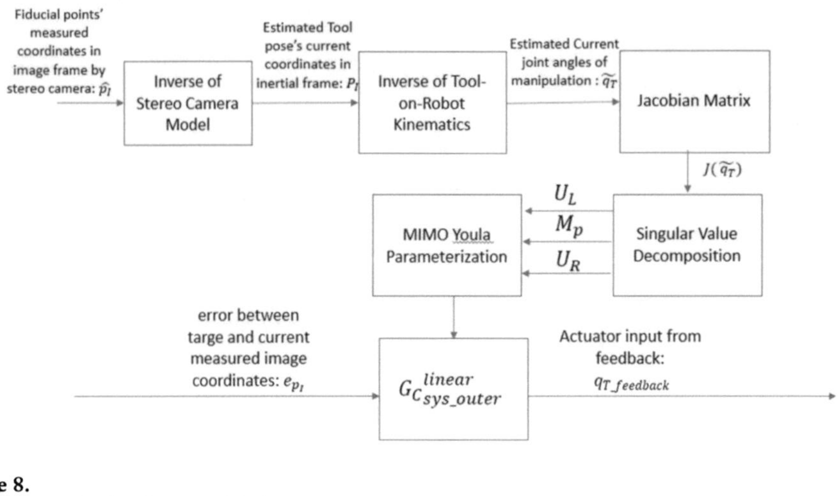

Figure 8.
Block diagram of adaptive control design.

The first step is to estimate current joint angles $\tilde{q_T}$ from current measured images coordinates $\widehat{p_I}$. The mathematic models of stereo camera and robot kinematics have been given in the above sections, and a combined model expression is defined in Eq. (30). Therefore, the mathematical function of the inverse model can be derived and expressed as:

$$\tilde{q_T} = \mathcal{F}^{-1}(\widehat{p_I}) \tag{42}$$

Where $\mathcal{F}^{-1}$ is the inverse function of Eq. (30). Given the estimated current angle $\tilde{q_T}$, we can calculate the Jacobian matrix of the nonlinear model at current time. By obtaining left and right unimodular matrices and Smith-McMillan form from singular value decomposition, the current linear controller $Gc_{sys\_outer}^{linear}$ can be built by steps Eqs. (30)–(41).

In this way, the adaptive outer-loop controller is computed online based on current linearized plane model, which is evaluated at current estimated robot's joint angles. The mapping between nonlinear plant model and linearized model is complete in all joint angles' configurations, thus the adaptive feedback controller is robust in the entire range of operations.

4.3 Feedforward control

Feedforward controller generates a target rotational angle $q_{T\_feedforward}$ based on the tool's coordinates at the target location. As shown in the Figure below, Feedforward controller contains two inverse processes of the original plant in the development. The first diagram illustrates the inverse process of the tool-on-robot kinematics model and the second diagram $T_{forward}$ shows the inverse process of inner-loop closed transfer function $T_{inner-loop}$.

Eq. (10) provides the mathematical expression of the tool-on-robot kinematics model. The inverse of Eq. (10) gives an expression of $\breve{q_T}$ from $\breve{P_I}\ \{O\}$ and $\breve{P_I}\ \{E\}$.We can define the inverse of tool-on-robot kinematics as follows:

$$\breve{q_T} = \mathcal{H}\left(\breve{P_I}\ \{O\}, \breve{P_I}\ \{E\}\right) \tag{43}$$

This process is so-called inverse kinematics of the robot manipulator, which finds the joint configurations given the coordinates of points in both the bottom frames and the end-effector frame.

The $T_{forward}$ can be designed as,

$$T_{forward} = \frac{1}{T_{inner-closed}} \frac{1}{\left(\tau_{forward}s + 1\right)^2} = \frac{(\tau_{in}s + 1)^3}{(3\tau_{in}s + 1)} \frac{1}{\left(\tau_{forward}s + 1\right)^2} \cdot I_{6\times 6} \tag{44}$$

The double poles s = $-1/\tau_{forward}$ are added to make $T_{forward}$ proper. Choose $\tau_{forward}$ so that the added double poles are 10 times larger than the bandwidth of the original improper $T_{forward}$. In other words, $\tau_{forward}$ is chosen as (**Figure 9**),

$$\tau_{forward} = 0.1\tau_{in} \tag{45}$$

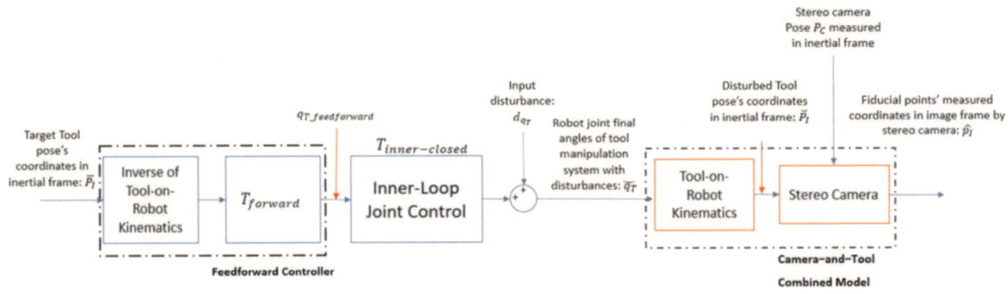

Figure 9.
Block diagram of feedforward control design.

4.4 Feature estimation

As the camera is static (not moving with the robot arm), the tool pose cannot be recognized and measured visually if it is outside the camera range of view. To tackle this problem, the 2D feature (image coordinates of the tool points) can be estimated from the same combined model in Eq. (10) with the joint angles q_T as input is shown in **Figure 3**:

$$\tilde{p}_I = \mathcal{F}(q_T) \tag{46}$$

5. Simulation results

5.1 Response of tool's pose in the control system

In this section, we will simulate the closed adaptive control loop with different scenarios to check the performance and stability of the proposed controller. For all scenarios, the tool is guided to a particular target position measured in the base inertial frame {O}: $\overline{P_I} = [-1m, 0.2m, 0.3m]^T$, where m represents meters and the tool should be aligned vertically so that its rotational matrix equals to $[\overline{n}, \overline{s}, \overline{a}] = \begin{bmatrix} 0 & -1 & 0 \\ -1 & 0 & 0 \\ 0 & 0 & -1 \end{bmatrix}$, where $\overline{n}, \overline{s}, \overline{a}$ represent respectively the end-effector's directional unit vector of the yaw, pitch, and roll in the inertial frame {O}. All scenarios are simulated with bandwidth of the inner-loop as 100 rad/s and the bandwidth of the outer-loop as 10 rad/s.

In the following three plots, the blue lines represent the responses of the end-effector position, and the red lines are the targets. In the trajectory plots, the circles represent the starting position of the end-effector, and the stars represent the target position of the end-effector. Black arrows are orientation direction of trajectory points that change over time, and red arrows are the target direction of the end-effector.

Three simulation scenarios are illustrated in **Figures 10–12**, respectively. In the first, the tool (end-effector) starts at the pose: $P_I^o = [1.404, 0.228m, 1.171m]^T$, and $[n_0, s_0, a_0] = \begin{bmatrix} -0.4893 & -0.0262 & 0.8717 \\ 0.2427 & 0.9560 & 0.1650 \\ -0.8377 & 0.2932 & -0.4614 \end{bmatrix}$, where two fiducial points are in side

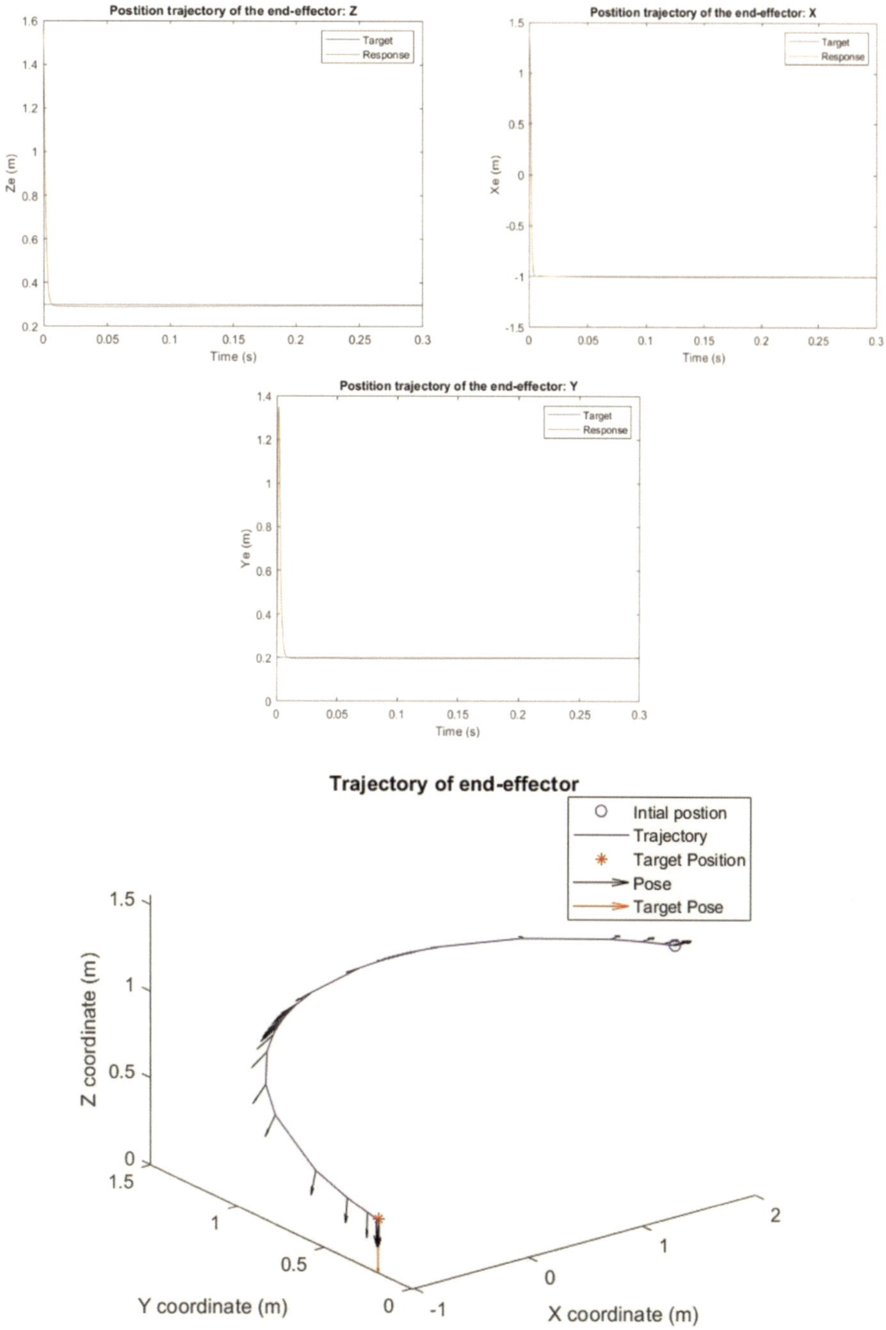

Figure 10.
Response of the tool's (end-effector's) pose in scenario 1: Start in the field view of camera with disturbances.

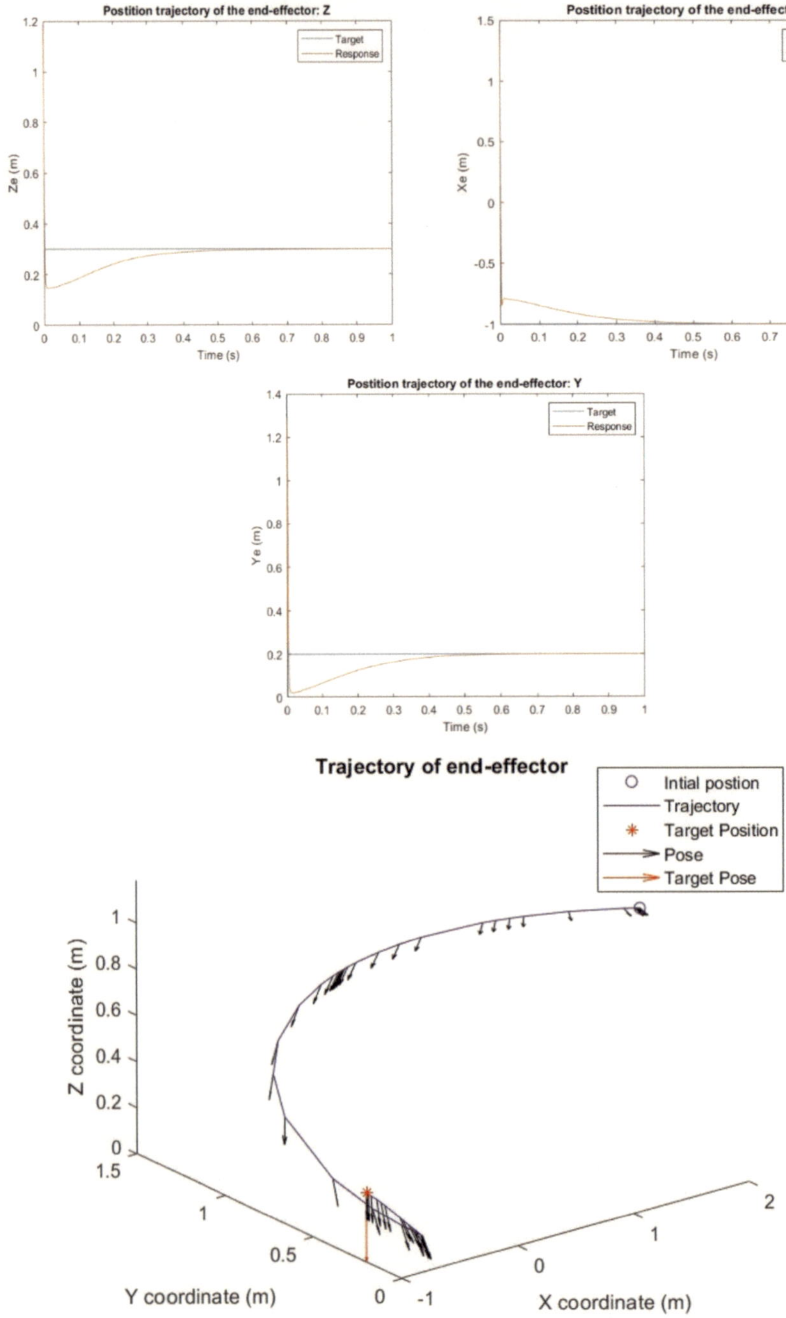

Figure 11.
Response of the tool's (end-effector's) pose in scenario 2: Start outside field view of camera with no disturbances.

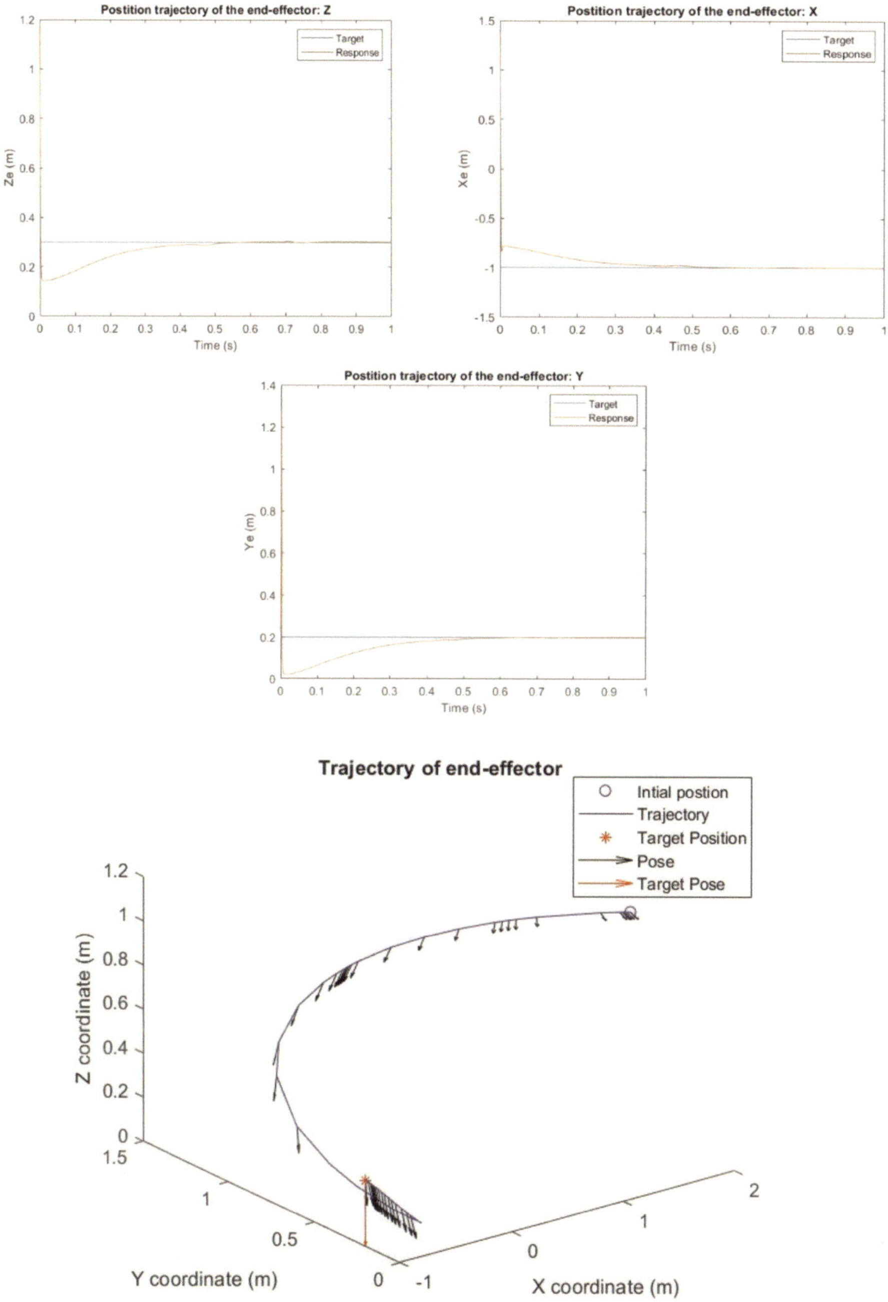

Figure 12.
Response of the tool's (end-effector's) pose in scenario 3: Start outside field view of camera with disturbances.

view of the camera. Therefore, their coordinates can be detected and measured during the entire control period. In the second and third scenario, the tool (end-effector) starts at the pose: $P_I^o = [1.285m, 0m, 1.57m]^T$, and $[n_0, s_0, a_0] = \begin{bmatrix} 0 & 0 & 1 \\ 0 & 1 & 0 \\ -1 & 0 & 0 \end{bmatrix}$, where two fiducial points are not inside view of the camera at beginning. Therefore, their coordinates must be estimated at the initial stage of the control loop. Disturbances of joint angles are added to the first and the third scenarios, as $d_q = [0.1^\circ, 0.5^\circ, 0.2^\circ, 0.3^\circ, -0.1^\circ, 0.3^\circ]$ to test the controller robustness against these input disturbances.

The response in **Figures 10** and **12** illustrate that the designed feedforward-feedback adaptive controller can reach the target in a stable manner even in the presence of input disturbances. Comparing orientation trajectories between **Figures 11** and **12**, disturbances add instability in the transient response, but the control system is able to stabilize it in the steady state response. When the tool's coordinates are measurable by the camera (as shown in **Figure 10**), the feedforward and feedback controller work together to guide the tool to its target pose fast and precisely, which is indicated as a response within 0.3 seconds and small overshoots in transient response. However, when the tool is outside the view of the camera (as shown in **Figures 11** and **12**), the controlled system still converges fast (less than a second) but generates large overshoots in some of the position responses. The large overshoots may come from the accumulated disturbances that cannot be eliminated by the feedforward control without the intervention of the feedback control. The issue of the overshoots will be investigated in future research and can be dealt with either by upgrading the camera with a wider range of view or by using a switching algorithm, such as switching from a feedforward to a feedback controller, rather than the use of a continuous feedforward-feedback controller.

5.2 Robustness analysis of model uncertainties

In this section, we study the impact of varying some parameters on the steady-state error of the end-effector X-coordinate position. The length of link 2 and link 4 of the robot manipulator L_2 and L_4 are chosen to vary as uncertainties because they have the two largest numerical dimensions among all geometric parameters of the robot manipulator.

Figure 13 shows the robustness test taken in simulation scenario 1. All points in the variation range have a steady state error which is less than 1%. We can use this as a standard to show the robust performance of the controller against model variation. From the plot, we can summarize that the designed control system is robust to model variation of the two parameters varying individually between 50 and 110%.

6. Conclusion

This paper proposes a new feedforward-feedback adaptive control algorithm based on image-based visual servoing technique. The core of this method is to design a feedforward controller and feature estimation loop so that the tool's pose can be stabilized when it is outside the view of the camera. Furthermore, an adaptive Youla parameterization-based controller is developed in the feedback loop to ensure stability

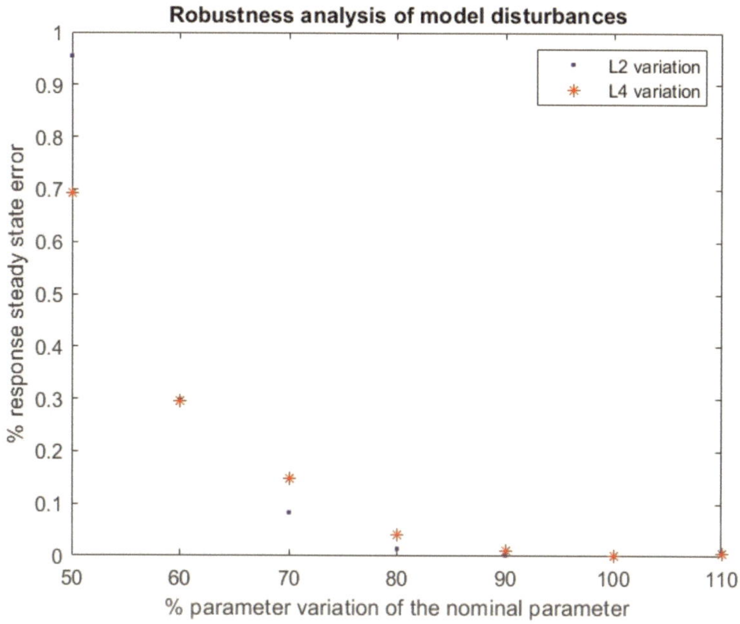

Figure 13.
Robustness analysis of model uncertainties.

and performance for the entire range of operations of the robot manipulator in the workspace. Various scenarios of the fastening and unfastening alignment problem have been simulated and robustness against input disturbances and model uncertainties have been examined. In summary, our control design has shown fast and robust performance in different scenarios so it has a great potential for implementation in real-world applications in automated manufacturing fields.

A. Appendix

See **Table 2**.

Parameters (Robot manipulator)	Value
Length of Link 1: L_1	0.495 m
Length of Link 2: L_2	0.9 m
Length of Link 3: L_3	0.175 m
Length of Link 4: L_4	0.96 m
Length of Link 1 offset: a_1	0.175 m

Parameters (Robot manipulator)	Value
Spherical wrist: L_t	0.135 m
Tool length of screwdriver:	0.127 m
Parameters (Zed 2 Stereo Camera)	*Value*
Focus length: f	2.8 mm
Baseline: b	120 mm
Angle of view in width: α	86.09°
Angle of view in height: β	55.35°

Table 2.
Parameter values of ABB IRB 4600 robot manipulator [31] and Zed 2 Stereo camera [33].

Author details

Rongfei Li and Francis F. Assadian*
University of California, Davis, California, USA

*Address all correspondence to: fassadian@ucdavis.edu

References

[1] Qin F, Xu D, Zhang D, Pei W, Han X, Yu S. Automated hooking of biomedical microelectrode guided by intelligent microscopic vision. IEEE/ASME Transactions on Mechatronics. 2023; **28**(5):2786-2798. DOI: 10.1109/TMECH.2023.3248112

[2] Jung W, Jin L, Hyun J. Automatic Alignment Type Welding Apparatus and Welding Method Using the Above Auto-Type Welding Apparatus; 2020

[3] Guo P, Zhang Z, Shi L, Liu Y. A contour-guided pose alignment method based on Gaussian mixture model for precision assembly. Assembly Automation. 2021;**41**(3):401-411

[4] Chang W, Wu C-H. Automated USB peg-in-hole assembly employing visual servoing. In: Proceedings of the 2023 3rd International Conference on Robotics (ICCAR). Nagoya, Japan: IEEE; 2017. pp. 352-355

[5] Xu J, Liu K, Pei Y, Yang C, Cheng Y, Liu Z. A noncontact control strategy for circular peg-in-hole assembly guided by the 6-DOF robot based on hybrid vision. IEEE Transactions on Instrumentation and Measurement. 2022;**71**:Art. no. 3509815

[6] Zhu W, Liu H, Ke Y. Sensor-based control using an image point and distance features for rivet-in-hole insertion. IEEE Transactions on Industrial Electronics. 2020;**67**(6): 4692-4699

[7] Chaumette F, Hutchinson S. Visual servo control part I: Basic approaches. IEEE Robotics and Automation Magazine. 2006;**13**:82-90

[8] Chaumette F. Image moments: A general and useful set of features for visual servoing. IEEE Transactions on Robotics. 2004;**20**:713-723

[9] Collewet C, Marchand E, Chaumette F. Visual servoing set free from image processing. In: IEEE International Conference on Robotics and Automation. Pasadena, California, USA: IEEE; 2008. pp. 81-86

[10] Cervera E, del Pobil AP, Berry F, Martinet P. Improving image-based visual servoing with three-dimensional features. The International Journal of Robotics Research. 2003;**22**(10–11): 821-839

[11] Kanatani K. Linear algebra method for pose optimization computation. In: Sergiyenko O, Flores-Fuentes W, Mercorelli P, editors. Machine Vision and Navigation. New York: Springer; 2020. pp. 293-319

[12] Feddema J, Lee CSG, Mitchell O. Model-based visual feedback control for a hand-eye coordinated robotic system. Computer. 1992;**25**(8):21-31

[13] Mezouar Y, Chaumette F. Optimal camera trajectory with image-based control. The International Journal of Robotics Research. 2003;**22**(10):781-804

[14] Janabi-Sharifi F, Deng L, Wilson W. Comparison of basic visual servoing methods. IEEE/ASME Transactions on Mechatronics. 2011;**16**(5):967-983

[15] Nematollahi E, Janabi-Sharifi F. Generalizations to control laws of image-based visual servoing. International Journal of Optomechatronics. 2009;**3**(3): 167-186

[16] Wang H, Liu Y-H, Chen W. Uncalibrated visual tracking control without visual velocity. IEEE Transactions on Control Systems Technology. 2010;**18**(6):1359-1370

[17] Cai C, Dean-León E, Somani N, Knoll A. 3D image-based dynamic visual servoing with uncalibrated stereo cameras. In: IEEE ISR 2013. Seoul, Korea (South): IEEE; 2013. pp. 1-6. DOI: 10.1109/ISR.2013.6695650

[18] Ma Y, Liu X, Zhang J, Xu D, Zhang D, Wu W. Robotic grasping and alignment for small size components assembly based on visual servoing. International Journal of Advanced Manufacturing Technology. 2020;**106**(11–12):4827-4843

[19] Hao T, Xu D, Qin F. Image-based visual Servoing for position alignment with orthogonal binocular vision. IEEE Transactions on Instrumentation and Measurement. 2023;**72**:1, Art no. 5019010-10. DOI: 10.1109/TIM.2023.3289560

[20] Chaumette F, Hutchinson S. Visual Servoing and visual tracking. In: Siciliano B, Oussama K, editors. Handbook of Robotics. Berlin Heidelberg, Germany: Springer-Verlag; 2008. pp. 563-583. DOI: 10.1007/978-3-540-30301-5.ch25

[21] Wilson WJ, Hulls CCW, Bell GS. Relative end-effector control using Cartesian position based visual servoing. IEEE Transactions on Robotics and Automation. 1996;**12**(5):684-696. DOI: 10.1109/70.538974

[22] Hashimoto K, Ebine T, Kimura H. Dynamic visual feedback control for a hand-eye manipulator. In: Proceedings of the 1992 lEEE/RSJ International Conference on Intelligent Robots and Systems. Vol. 3. Raleigh, NC, USA: IEEE; 1992. pp. 1863-1868

[23] Zergeroglu E, Dawson D, de Queiroz M, Nagarkatti S. Robust visual-servo control of robot manipulators in the presence of uncertainty. In: IEEE Conference on Decision and Control. Vol. 4. Phoenix, AZ, USA: IEEE; 1999. pp. 4137-4142

[24] Youla D, Jabr H, Bongiorno J. Modern wiener-Hopf Design of Optimal Controllers-Part II: The multivariable case. IEEE Transactions on Automatic Control. 1976;**21**(3):319-338. DOI: 10.1109/TAC.1976.1101223

[25] Liu S, Xu D, Liu F, Zhang D, Zhang Z. Relative pose estimation for alignment of long cylindrical components based on microscopic vision. IEEE/ASME Transactions on Mechatronics. 2016;**21**(3):1388-1398

[26] Liu Y et al. An adaptive controller for image-based visual servoing of robot manipulators. In: 2010 8th World Congress on Intelligent Control and Automation. Jinan, China: IEEE; 2010. DOI: 10.1109/wcica.2010.5554505

[27] Slotine JJ, Li W. On the adaptive control of robot manipulators. International Journal of Robotics Research. 1987;**6**:49-59

[28] Liu A et al. Resilient adaptive trajectory tracking control for uncalibrated visual servoing systems with unknown actuator failures. Journal of the Franklin Institute. 2024;**361**(1): 526-542. DOI: 10.1016/j.jfranklin.2023.12.011

[29] Qiu Z et al. Adaptive neural network control for image-based visual servoing of robot manipulators. IET Control Theory and Applications. 2022;**16**(4): 443-453. DOI: 10.1049/cth2.12238

[30] Illingworth J, Kittler J. A survey of the Hough transform. Computer Vision, Graphics, and Image Processing. 1988; **44**(1):87-116

[31] Anonymous. ABB IRB 4600–40/2.55 Product Manual [Internet]. 2013. Available from: https://www.manualslib.com/manual/1449302/Abb-Irb-4600-40-2-55.html#manual [Accessed: December 01, 2021]

[32] Denavit J, Hartenberg RS. A kinematic notation for lower-pair mechanisms based on matrices. American Society of Mechanical Engineers. 1955, 1955;**23**(2):215-221. DOI: 10.1115/1.4011045

[33] Anonymous. ZED 2 Camera and SDK Overview [Internet]. STEREOLABS; Available from: https://cdn2.stereolabs.com/assets/datasheets/zed2-camera-datasheet.pdf

[34] Li R, Assadian F. Role of uncertainty in model development and control Design for a Manufacturing Process. In: Majid T, Pengzhong L, Liang L, editors. Production Engineering and Robust Control. London: IntechOpen; 2022. pp. 137-167. DOI: 10.5772/intechopen.101291

[35] Assadian F, Mallon K. Robust Control: Youla Parameterization Approach. New Jersy: Willey; 2022